The Shanghai Maths Project

Maths Project

For the English National Curriculum

一课一练

Practice Book 6B

Series Editor: Professor Lianghuo Fan

UK Curriculum Consultant: Paul Broadbent

Collins

William Collins' dream of knowledge for all began with the publication of his first book in 1819.

A self-educated mill worker, he not only enriched millions of lives, but also founded a flourishing publishing house. Today, staying true to this spirit, Collins books are packed with inspiration, innovation and practical expertise. They place you at the centre of a world of possibility and give you exactly what you need to explore it.

Collins. Freedom to teach.

MIX
Paper from responsible sources
www.fsc.org   **FSC™ C007454**

This book is produced from independently certified FSC paper to ensure responsible forest management.

For more information visit:
www.harpercollins.co.uk/green

Published by Collins
An imprint of HarperCollins*Publishers*
The News Building
1 London Bridge Street
London
SE1 9GF

Browse the complete Collins catalogue at
www.collins.co.uk

HarperCollins *Publishers*
1st Floor
Watermarque Building
Ringsend Road
Dublin 4
Ireland

The Shanghai Maths Project (for the English National Curriculum) is a collaborative effort between HarperCollins, East China Normal University Press Ltd. and Professor Lianghuo Fan and his team. Based on the latest edition of the award-winning series of learning resource books, *One Lesson, One Exercise*, by East China Normal University Press Ltd. in Chinese, the series of Practice Books is published by HarperCollins after adaptation following the English National Curriculum.

Practice Book Year 6B has been translated and developed by Professor Lianghuo Fan with the assistance of Ellen Chen, Ming Ni, Huiping Xu and Dr Jane Hui-Chuan Li, with Paul Broadbent as UK Curriculum Consultant.

Series Editor: Professor Lianghuo Fan
UK Curriculum Consultant: Paul Broadbent
Publishing Manager: Fiona McGlade and Lizzie Catford
In-house Editor: Mike Appleton
In-house Editorial Assistant: August Stevens
Project Manager: Karen Williams
Copy Editors: Catherine Dakin and Tracy Thomas
Proofreaders: Tracy Thomas, Steven Matchett and Amanda Dickson
Cover design: Kevin Robbins and East China Normal University Press Ltd.
Cover artwork: Daniela Geremia
Internal design: 2Hoots Publishing Services Ltd
Typesetting: Ken Vail Graphic Design Ltd
Illustrations: Ken Vail Graphic Design Ltd
Production: Sarah Burke
Printed and Bound in the UK using 100% Renewable Electricity at CPI Group (UK) Ltd

Contents

Chapter 6 Operations with fractions

6.1 Fractions and division

Learning objective Understand the relationship between fractions and division

Basic questions

1 Multiple choice questions. (For each question, choose the correct answer and write the letter in the box.)

(a) If 5 kilograms of coal can generate 9 kilowatt-hours (kWh) of electricity, then to generate 1 kilowatt-hour of electricity, there needs to be ☐ of coal.

A. $\frac{1}{9}$ kg **B.** $\frac{1}{5}$ kg **C.** $\frac{9}{5}$ kg **D.** $\frac{5}{9}$ kg

(b) Ben is reading a 50-page detective novel. If he read 10 pages on the first day and 9 pages on the second day, then he read ☐ of the book on the second day.

A. $\frac{19}{50}$ **B.** $\frac{9}{10}$ **C.** $\frac{9}{50}$ **D.** $\frac{10}{9}$

2 Fill in the spaces.

(a) $\frac{9}{17}$ is ☐ lots of $\frac{1}{17}$. 9 lots of $\frac{1}{19}$ is ☐. There are ☐ lots of $\frac{1}{47}$ in $\frac{9}{47}$.

(b) Use fractions to express the answers to the divisions below.

(i) $7 \div 16 =$ ☐ (ii) $16 \div 7 =$ ☐

(iii) $6 \div 7 =$ ☐ (iv) $11 \div 6 =$ ☐

(c) Express these fractions as division sentences with two integers.

(i) $\frac{3}{4}$ = _____

(ii) $\frac{9}{7}$ = _____

(iii) $\frac{13}{6}$ = _____

(iv) $\frac{6}{17}$ = _____

(d) If you divide a basket of apples weighing 15 kg into 4 equal parts, each part is ☐ kg. Each part is ☐ of 15 kg. (Fill in with a fraction.)

(e) 17 minutes = ☐ hour. (Fill in with a fraction.)

(f) Jimmy spent 19 minutes completing 12 questions; he spent ☐ minutes on each question on average.

(Fill in with a fraction.)

3 Fill in each box below the number line with a suitable fraction.

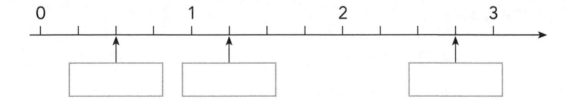

4 Mark the fractions $\frac{2}{3}$ and $\frac{4}{3}$ on the number line.

5 Look at each figure below. What fraction of each figure is shaded?

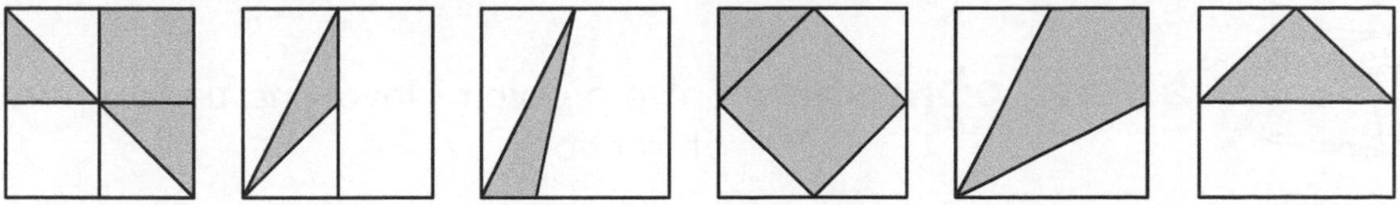

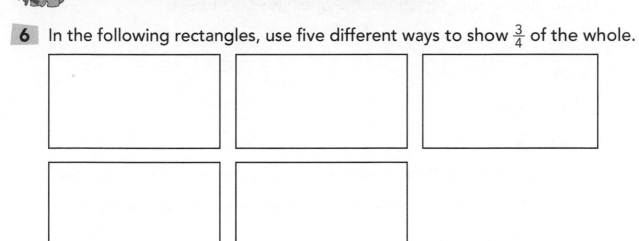

Challenge and extension question

6 In the following rectangles, use five different ways to show $\frac{3}{4}$ of the whole.

6.2 Basic properties of fractions (1)

Learning objective Solve problems involving the properties of fractions

Basic questions

1 Multiple choice questions. (For each question, choose the correct answer and write the letter in the box.)

(a) In the following fractions, the one not equal to $\frac{8}{12}$ is ☐ .

    A. $\frac{2}{3}$

    B. $\frac{4}{6}$

    C. $\frac{10}{18}$

    D. $\frac{30}{45}$

(b) Jim completed 10 questions in quarter of an hour. To express the amount of time Jim took, the incorrect one of the following is ☐ .

    A. $\frac{15}{60}$ hours

    B. $\frac{1}{4}$ hours

    C. 0.15 hours

    D. 15 minutes

(c) In the following statements, the correct one is ☐ .

    A. When both the numerator and denominator are multiplied by a number, the value of the fraction remains unchanged.

    B. When the numerator of a fraction is increased to 3 times the original numerator, and its denominator is decreased to $\frac{1}{3}$ of the original denominator, the value of the fraction is unchanged.

    C. 24 minutes is $\frac{2}{5}$ hours.

    D. When both the numerator and denominator are added to a number, the value of the fraction remains unchanged.

2 Fill in the missing numbers.

   (a) Simplify $\frac{6}{12}$ to a fraction with 4 as the denominator.

      The fraction is ◻.

   (b) Simplify $\frac{6}{12}$ to a fraction with 3 as the numerator.

      The fraction is ◻.

   (c) Given $\frac{1}{4} = \frac{2}{8} = \frac{4}{16}$, we know that there are ◻ lots of $\frac{1}{8}$ in $\frac{1}{4}$ and ◻ lots of $\frac{1}{16}$ in $\frac{1}{4}$. Similarly, there are ◻ lots of $\frac{1}{18}$ in $\frac{1}{2}$.

   (d) Fill in each box with a suitable number.

      (i) $\frac{\square}{25} = \frac{16}{\square} = 0.8 = 24 \div \square$

      (ii) $\frac{2+\square}{3+12} = \frac{2}{3} = \frac{2+4}{3+\square}$

   (e) If 8 is added to the numerator of $\frac{2}{15}$, and the value of the fraction remains unchanged, ◻ should be added to the denominator.

3 Write all the fractions with the denominators less than 30 that have the same value as $\frac{12}{30}$.

_____

4 Use at least two methods to increase the value of $\frac{12}{30}$, so the value of the new fraction is 5 times the original one. Explain your reason.

_____

_____

Challenge and extension question

5 If positive integers a and b satisfy $\frac{b}{a^2} = \frac{7}{18}$, find the least values of a and b.

6.3 Basic properties of fractions (2)

 Learning objective Use common factors to simplify fractions

 Basic questions

1. Multiple choice questions. (For each question, choose the correct answer and write the letter in the box.)

   (a) In the following fractions, ☐ is not in its simplest form.

   A. $\frac{5}{8}$ **B.** $\frac{5}{16}$ **C.** $\frac{17}{19}$ **D.** $\frac{13}{91}$

   (b) Ben studied for 3 hours in the morning. It was ☐ of a day of 24 hours.

   A. $\frac{1}{24}$ **B.** $\frac{1}{12}$ **C.** $\frac{1}{8}$ **D.** $\frac{1}{6}$

   (c) In $\frac{3}{7}$, $\frac{21}{24}$, $\frac{19}{38}$, $\frac{13}{52}$ and $\frac{7}{9}$, there are ☐ fractions in their simplest form.

   A. 1   **B.** 2   **C.** 3   **D.** 4

2. Complete each statement.

   (a) When both the numerator and the denominator of a fraction have no

   _____ factor except 1, it is called 'the fraction in its simplest form'. Dividing both the numerator and the denominator of a fraction

   by all their _____ factors, we can simplify the fraction to its simplest form.

   (b) The denominator of a fraction is 45. After simplification, it is $\frac{2}{9}$.

   The fraction is ☐.

   (c) Fill in the box with a fraction in its simplest form: 2250 g = ☐ kg

   (d) Three years ago, Jack was 12 years old and his mum was 42 years old.

   Now Jack's age is ☐ of his mum's. (Fill in with a fraction.)

3 Simplify each fraction to its simplest form. The first one has been done for you.

(a) $\frac{6}{9}$

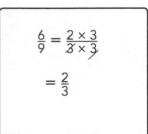

$$\frac{6}{9} = \frac{2 \times 3}{3 \times 3}$$
$$= \frac{2}{3}$$

(b) $\frac{25}{10}$

(c) $\frac{48}{42}$

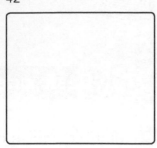

(d) $\frac{125}{1000}$

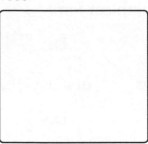

(e) $\frac{26}{39}$

(f) $\frac{74}{111}$

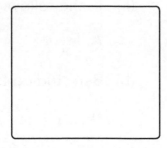

4 Find the value of x using basic properties of fractions.

(a) $\frac{48}{x} = \frac{36}{24}$

(b) $\frac{x}{24} = \frac{42}{18}$

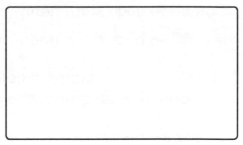

5 The common factor of the numerator and denominator of a fraction is 13 and it is $\frac{2}{3}$ after simplification. Find the fraction.

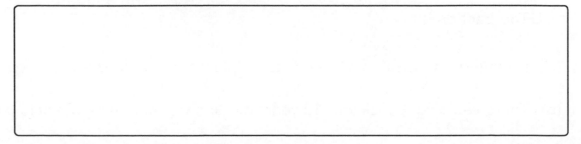

6 The sum of the numerator and denominator of a fraction is 156 and it is $\frac{5}{8}$ in its simplest form. Find the fraction.

6.4 Basic properties of fractions (3)

Learning objective Solve problems involving equivalent fractions

Basic questions

1 Multiple choice questions. (For each question, choose the correct answer and write the letter in the box.)

(a) 20 g of sugar is mixed with 100 g of flour. The sugar takes up ☐ of the resulting mixture.

    A. $\frac{1}{5}$ **B.** $\frac{1}{6}$ **C.** 5   **D.** 6

(b) Mr Rob planned to make 100 spare parts in a day. Given that he made 40 spare parts in the morning and 50 in the afternoon, he finished ☐ of his plan.

    A. $\frac{2}{5}$ **B.** $\frac{9}{10}$ **C.** $\frac{1}{2}$ **D.** $\frac{10}{9}$

(c) If the original price of an item was £250 and it is now selling for £50 less, then the sale price is ☐ of the original price.

    A. $\frac{1}{5}$ **B.** $\frac{4}{5}$ **C.** $\frac{5}{4}$ **D.** $\frac{1}{4}$

2 Complete each statement using a fraction.

(a) Ted received £200 as a Christmas gift. He donated £100 to a charity.

The money he donated is ☐ of the gift money he received.

(b) Given that the height of the Shard in London is 306 m and that of Heron Tower is 230 m, then the height of Heron Tower is ☐ of the height of the Shard.

(c) Given that there are 150 literature books and 650 science books on the bookshelves in a library, then the science books take up ☐ of all the books on the bookshelves.

(d) Mrs Evans was working on a project that was scheduled to be completed in 10 hours. After working very efficiently, she completed it 2 hours ahead of schedule. From 7:30 a.m. to 2:30 p.m., Mrs Evans had completed ☐ of the project.

(e) If A is $\frac{1}{3}$ more than B, then B is ☐ less than A.

3 The table shows the results of a class Maths test. What fraction of the pupils in the class scored 80 marks and above? ☐

Score (marks)	below 60	60–79	80–100
Number of pupils	1	23	20

4 The line graph shows the production (in thousand pounds) of a toy factory in the first quarter of a year. Based on the data given in the graph, complete the table on the next page.

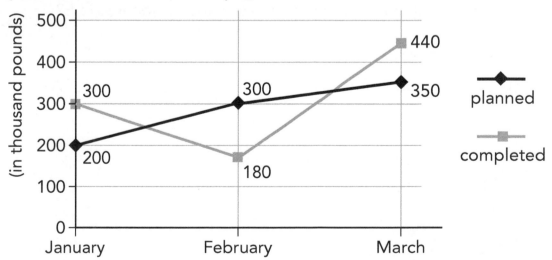

Production (in thousand pounds) of a toy factory in the first quarter

Items	Month			Total
	January	February	March	
planned	200			
completed	300			
completed production as a fraction of planned production	$\frac{3}{2}$			

Challenge and extension question

5 The table shows the results of a class Maths test. However, the cell for the number of pupils who scored 50 to 59 marks is illegible.

Score (marks)	above 90	80–89	70–79	60–69	50–59	below 50
Number of pupils	8	10	5	3		1

Pupils who scored above 90 marks received a grade of A, and $\frac{2}{15}$ of the class scoring below 60 marks received a grade of E.

(a) How many pupils in the class received a grade above E?

(b) How many pupils in the class scored 50 to 59 marks?

(c) What fraction of pupils in the class received a grade of A?

6.5 Comparison of fractions

Learning objective Compare the value of fractions using common denominators

Basic questions

1 Multiple choice questions. (For each question, choose the correct answer and write the letter in the box.)

(a) In the following fractions, the fraction greater than $\frac{1}{3}$ but less than $\frac{1}{2}$ is ☐.

    A. $\frac{5}{12}$ **B.** $\frac{4}{13}$ **C.** $\frac{7}{12}$ **D.** $\frac{6}{12}$

(b) In $\frac{1}{3}$, $\frac{2}{5}$, $\frac{3}{8}$ and $\frac{4}{11}$, there are ☐ numbers greater than $\frac{1}{4}$ but less than $\frac{1}{2}$.

    A. 1   **B.** 2   **C.** 3   **D.** 4

(c) Chris and Allen were tasked to type the same article. If Chris spent $\frac{16}{17}$ hours and Allen spent $\frac{8}{9}$ hours to complete the task, who typed faster? ☐

    A. Chris

    B. Allen

    C. The speeds were the same

    D. Can't compare

(d) In the following numbers, the number that is not a common denominator of $\frac{2}{3}$ and $\frac{3}{5}$ is ☐.

    A. 15

    B. 30

    C. 50

    D. 60

> A common multiple of the two denominators of two fractions is called a **common denominator** of the fractions.

(e) In the following numbers, the lowest common denominator of

$\frac{1}{4}$ and $\frac{5}{6}$ is ☐ .

A. 4

B. 6

C. 12

D. 24

The lowest value of all the common denominators is called the **lowest common denominator**.

2 Complete each statement.

(a) For fractions with the same denominator, the _____ the numerator, the greater the fraction. For fractions with the same numerator, the _____ the denominator, the greater the fraction.

(b) Compare the following fractions.

   (i) $\frac{5}{21}$ _____ $\frac{4}{21}$

   (ii) $\frac{3}{5}$ _____ $\frac{3}{8}$

   (iii) $2\frac{2}{3}$ _____ $3\frac{1}{3}$

(c) Write three common denominators of $\frac{5}{6}$ and $\frac{3}{8}$: _____

(d) Find the lowest common denominator for each pair of fractions.

   (i) The lowest common denominator of $\frac{4}{3}$ and $\frac{3}{4}$ is _____ .

   (ii) The lowest common denominator of $\frac{2}{25}$ and $\frac{2}{12}$

   is _____ .

   (iii) The lowest common denominator of $\frac{1}{6}$ and $\frac{7}{20}$ is _____ .

(e) Fractions in the simplest form with the denominator 14 and less than $\frac{6}{7}$ are _____.

(f) There are _____ fractions that are less than $\frac{8}{9}$ but greater than $\frac{7}{8}$.

3 Put the fractions $\frac{5}{8}$, $\frac{7}{10}$, $\frac{2}{5}$, $\frac{3}{6}$ and $\frac{6}{9}$ in order, from the least to the greatest.

4 A fraction is greater than $\frac{3}{4}$ but less than $\frac{6}{7}$; it is in its simplest form with a denominator of 28. Find the fraction.

5 Team A, Team B and Team C were assigned to complete the same amount of work. Team A took $\frac{10}{7}$ hours, Team B took 1.4 hours and Team C took 1 hour and 27 minutes to complete the work. Which team had the highest efficiency?

Challenge and extension question

6 Look at the following diagrams and think carefully. Then answer the questions below.

$\frac{1}{2}$ $\frac{1+1}{2+1} = \frac{2}{3}$ $\frac{2+1}{3+1} = \frac{3}{4}$ $\frac{3+1}{4+1} = \frac{4}{5}$ $\frac{4+1}{5+1} = \frac{5}{6}$

(a) Use < to put the five fractions $\frac{1}{2}$, $\frac{2}{3}$, $\frac{3}{4}$, $\frac{4}{5}$ and $\frac{5}{6}$ in order, from the least to the greatest.

_____

(b) What pattern did you find? Which fraction is greater: $\frac{2011}{2012}$ or $\frac{2012}{2013}$?

_____

6.6 Addition and subtraction of fractions (1)

Learning objective Add and subtract fractions with different denominators

Basic questions

1 Complete each statement.

(a) When adding or subtracting fractions with the same denominator, we

_____ the numerators and keep the denominator unchanged.

When adding or subtracting fractions with different denominators,

we first find a common _____ for the denominators. Then we convert the fractions so that they all have that value for the denominator. It is more efficient to use

the _____ common denominator. It is possible, then, to add or subtract numerators because all the denominators are the same.

(b) Calculate the answers.

(i) $\frac{3}{8} + \frac{5}{8} =$ _____

(ii) $\frac{13}{15} - \frac{7}{15} =$ _____

(iii) $\frac{1}{4} + \frac{1}{6} =$ _____

(iv) $\frac{1}{4} - \frac{1}{6} =$ _____

(c) Fill in each box with a suitable number.

(i) $\boxed{} + \frac{4}{15} = \frac{7}{15}$

(ii) $\frac{16}{19} - \boxed{} = \frac{5}{19}$

(iii) $\frac{7}{8} - \boxed{} = \frac{1}{56}$

(iv) $\boxed{} + \frac{5}{23} + \frac{12}{23} = \frac{21}{23}$

(d) Look at these figures and write a fraction in each box below.

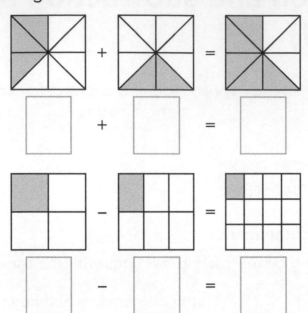

$$\boxed{} + \boxed{} = \boxed{}$$

$$\boxed{} - \boxed{} = \boxed{}$$

(e) Fill in the cells in each right-hand column with suitable numbers.

(i)

Input
| $\frac{7}{10}$ |
| $\frac{5}{6}$ |
| $\frac{2}{3}$ |
| $\frac{11}{15}$ |

$+ \frac{2}{5}$ →

Output

(ii)

Input
| $\frac{1}{3}$ |
| $\frac{2}{5}$ |
| $\frac{5}{8}$ |
| $\frac{9}{14}$ |

$- \frac{1}{7}$ →

Output

2 Calculate the answers.

(a) $\frac{7}{15} - \frac{7}{20}$

(b) $\frac{3}{5} + \frac{7}{20} - \frac{3}{4}$

(c) $\frac{4}{7} - \frac{2}{5} + \frac{2}{21}$

3 Write down the number sentences and the output when (i) $x = \frac{1}{4}$ and (ii) $x = \frac{2}{3}$ respectively.

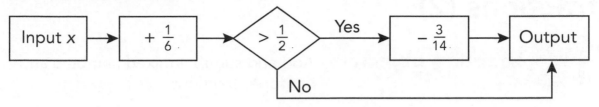

(i) _____

(ii) _____

Challenge and extension question

4 Write the fractions $\frac{11}{18}$ and $\frac{4}{5}$ into the following flowchart, and put the result in the output brackets.

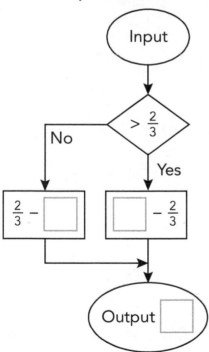

6.7 Addition and subtraction of fractions (2)

Learning objective Add and subtract mixed numbers and improper fractions

Basic questions

1 Multiple choice questions. (For each question, choose the correct answer and write the letter in the box.)

(a) Fraction $\frac{27}{5}$ is between ☐.

    A. 3 and 4   **B.** 5 and 6   **C.** 7 and 8   **D.** 9 and 10

(b) If $\frac{5}{x+2}$ is an improper fraction and x is a positive integer, then there

    are ☐ possible values of x.

    A. 1   **B.** 2   **C.** 3   **D.** 4

2 Complete each statement.

(a) Convert mixed numbers to improper fractions.

    (i) $3\frac{1}{3} =$ ☐   (ii) $3\frac{3}{8} =$ ☐   (iii) $5\frac{7}{10} =$ ☐

(b) Convert improper fractions to mixed numbers.

    (i) $\frac{34}{15} =$ ☐   (ii) $\frac{23}{7} =$ ☐   (iii) $\frac{145}{12} =$ ☐

(c) Fill in the space with a mixed number: 1 m 20 cm = _____ m.

(d) All the proper fractions in the simplest form with the same

    denominator of 18 are _____.

(e) If $\frac{a}{9}$ is a proper fraction and $\frac{a}{5}$ is an improper fraction, then the natural number a can be _____.

(f) If $\frac{m}{6}$ is a proper fraction in the simplest form, then the natural number m can be _____.

3 Calculate the answers.

(a) $12\frac{4}{5} + 9\frac{3}{10} - 4\frac{4}{15}$

(b) $11\frac{3}{14} - \left(\frac{3}{5} + 2\frac{5}{14}\right)$

4 Calculate the answers.

(a) $5\frac{1}{6} - \left(1\frac{1}{3} + 2\frac{2}{5}\right)$

(b) $6\frac{4}{18} - 3\frac{1}{6} + 2\frac{4}{9}$

5 Write the number sentences and then calculate the answers.

(a) $1\frac{7}{15}$ is added to the difference of $\frac{7}{10}$ and $\frac{7}{15}$. What is the sum?

(b) After a number is subtracted by $2\frac{5}{6}$ and then added by $1\frac{2}{3}$, it is equal to $3\frac{1}{2}$. Find the number.

Challenge and extension question

6 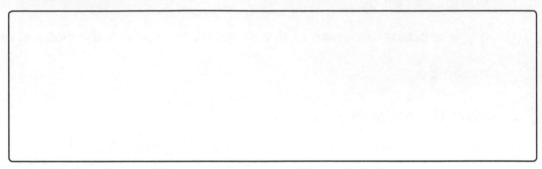 stands for a new operation, and the rule is:

a ▲ $b = (a + 3\frac{1}{2}) - (b + 2\frac{1}{2})$.

For example: 1 ▲ $2 = (1 + 3\frac{1}{2}) - (2 + 2\frac{1}{2}) = 0$.

Calculate the value of the following.

(a) $4\frac{5}{6}$ ▲ $5\frac{1}{6}$

(b) $10\frac{27}{28}$ ▲ $(7\frac{1}{14}$ ▲ $\frac{3}{28})$

6.8 Addition and subtraction of fractions (3)

Learning objective Solve missing number problems when adding and subtracting mixed numbers and improper fractions

Basic questions

1 Solve the equations for x. The first one has been done for you.

(a) $x - 4\frac{1}{3} = 3\frac{1}{4}$

$$x = 3\frac{1}{4} + 4\frac{1}{3}$$
$$= \left(3 + \frac{1}{4}\right) + \left(4 + \frac{1}{3}\right)$$
$$= (3 + 4) + \left(\frac{1}{4} + \frac{1}{3}\right)$$
$$= 7 + \left(\frac{3}{12} + \frac{4}{12}\right)$$
$$= 7\frac{7}{12}$$

(b) $x + 1\frac{16}{25} = 10$

(c) $4\frac{5}{8} + x = 7\frac{1}{2}$

(d) $10 - x = 9\frac{17}{25}$

(e) $x + 2\frac{8}{15} = 4\frac{1}{6}$

(f) $\frac{11}{20} + x = 1\frac{1}{20} + \frac{13}{20}$

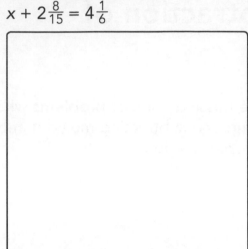

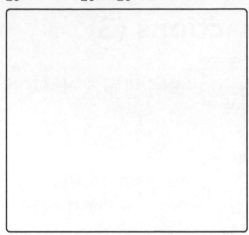

2 The difference of $5\frac{5}{14}$ minus a number is $2\frac{1}{5}$. Find the number.

3 The difference of the sum of $2\frac{2}{3}$ and a number minus $1\frac{4}{5}$ is $7\frac{9}{10}$. Find the number.

4 $1\frac{4}{5}$ is subtracted from the sum of $9\frac{1}{5}$ and $3\frac{7}{15}$. The difference is $\frac{2}{15}$ greater than a number. Find the number.

5 There are two buckets. Bucket A has $6\frac{1}{3}$ litres more water than Bucket B. If $2\frac{1}{4}$ litres of water from Bucket A is poured into Bucket B, how much more water would be in Bucket A than in Bucket B?

Challenge and extension question

6 An electrician planned to use a 40 m-long bundle of electrical wire to wire a new building. On the first day, he used $10\frac{2}{5}$ m of the wire, which was $\frac{3}{5}$ m less than that on the second day. The wire he used on the second day was $12\frac{1}{2}$ m less than that on the third day. Was the bundle of electrical wire long enough? Show your working.

6.9 Multiplication of fractions

Learning objective Multiply fractions and mixed numbers

Basic questions

1 Multiple choice questions. (For each question, choose the correct answer and write the letter in the box.)

(a) A 100 m-long wire first has $\frac{2}{5}$ of its length cut off and then a further 5 m cut off. It has ☐ m left.

    A. $94\frac{3}{5}$ m   **B.** 95 m   **C.** $99\frac{3}{5}$ m   **D.** 55 m

(b) To solve the following questions, the number sentence used not equal to $\frac{1}{2} \times \frac{4}{5}$ is ☐.

    A. A piece of string is $\frac{1}{2}$ m long. If $\frac{1}{5}$ of the string is cut off, what is the length of the remaining piece of string?

    B. The length of a rectangle is $\frac{4}{5}$ and the width is $\frac{1}{2}$. Find the area of the rectangle.

    C. When $\frac{1}{2}$ is divided by a number, the quotient is $\frac{4}{5}$. Find the number.

    D. When Peter rides a bicycle, his speed is twice his walking speed. If a journey takes him 1 hour by bike, find what fraction of the journey he can cover by walking in 48 minutes.

(c) Gina drew the following figures to explain that half of a half is a quarter, that is, $\frac{1}{2} \times \frac{1}{2} = \frac{1}{4}$. The incorrect one is ☐.

    A.   **B.**   **C.** 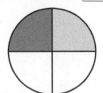   **D.**

2 Complete each statement.

(a) $\frac{3}{4}$ of $\frac{3}{5}$ tonnes is ☐ tonnes.

(b) $\frac{2}{5}$ of $\frac{2}{5}$ m is ☐ m.

(c) If 1 lesson period is $\frac{2}{3}$ hours, 6 lesson periods are ☐ hours.

(d) It takes a toy car $\frac{3}{5}$ minutes to complete 1 circuit. To complete

   5 circuits, it will take ☐ minutes.

(e) $\frac{2012}{2013} \times \frac{2011}{2012}$ _____ $\frac{2012}{2013}$ (Fill in with >, < or = .)

3 Calculate the answers. The first one has been done for you.

(a) $\frac{15}{28} \times \frac{7}{10}$

$$\frac{15}{28} \times \frac{7}{10}$$
$$= \frac{3 \times 5}{4 \times 7} \times \frac{7}{2 \times 5}$$
$$= \frac{3 \times 1}{4 \times 2}$$
$$= \frac{3}{8}$$

(b) $64 \times \frac{7}{8}$

(c) $\frac{1}{2} \times \frac{5}{7}$

(d) $\frac{13}{38} \times \frac{19}{26}$

(e) $\frac{2}{3} \times \frac{9}{14}$

(f) $\frac{16}{21} \times \frac{11}{12}$

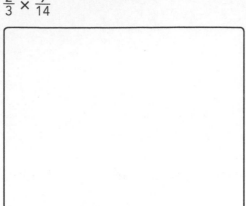

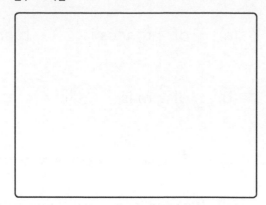

4 When a ball falls freely from a height, the height it bounces back to after it hits the ground is $\frac{2}{5}$ of its previous height. If a ball falls freely from a height of 35 metres, then what is the height after it bounces back the third time?

 Challenge and extension questions

5 Look at the diagram. If the rectangle represents 1, draw on the rectangle to show the meaning of $\frac{2}{3} \times \frac{1}{2}$. Show your drawing and explain.

6 Complete the calculation of $\frac{4}{15} \times 2\frac{1}{12}$.

Method 1:

$$\frac{4}{15} \times 2\frac{1}{12} = \frac{4}{15} \times \frac{\square}{12}$$

$$= \frac{4}{3 \times 5} \times \frac{5 \times \square}{3 \times 4}$$

$$= \frac{\square}{9}$$

Method 2:

$$\frac{4}{15} \times 2\frac{1}{12} = \frac{4}{15} \times \left(2 + \frac{\square}{12}\right)$$

$$= \frac{4}{15} \times 2 + \frac{4}{15} \times \frac{\square}{12}$$

$$= \frac{8}{15} + \frac{\square}{45}$$

$$= \frac{24}{45} + \frac{\square}{45}$$

$$= \boxed{}$$

Now, calculate the following. Show your working. Express your answers in mixed numbers.

(a) $\frac{7}{9} \times 3\frac{3}{14}$

(b) $1\frac{2}{3} \times 2\frac{3}{5}$

6.10 Dividing fractions by a whole number

Learning objective Divide proper fractions by whole numbers

Basic questions

1 Multiple choice questions. (For each question, choose the correct answer and write the letter in the box.)

(a) All the following questions except ☐ can be answered using the division: $\frac{1}{2} \div 10$.

A. Roy took 10 minutes to walk $\frac{1}{2}$ km. What was his speed per minute?

B. $\frac{1}{2}$ kg of sweets are shared equally by 10 children. How many kilograms of sweets should each child get?

C. Julie spent 10 days reading half a book. What fraction of the book did she read each day on average?

D. Ten pupils joined a reading group and half of them were girls. How many girls joined the reading group?

(b) Jim drew the following figures to explain the meaning of $\frac{1}{3} \div 2$.

The incorrect one is ☐.

A. **B.** 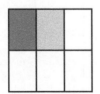 **C.** **D.**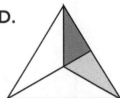

2 True or false? (Put a ✔ for true and a ✗ for false in each box.)

(a) A proper fraction divided by a whole number is also a proper fraction. □

(b) $\frac{1}{3} \div 2 = \frac{1}{3} \times \frac{1}{2}$ □

(c) In general, a fraction divided by a whole number m is equal to the fraction multiplied by $\frac{1}{m}$, or $\frac{b}{a} \div m = \frac{b}{a} \times \frac{1}{m}$ (a, b and m are all whole numbers, and a and m are not zero). □

(d) If a fraction is divided by a whole number m, then the numerator of the resulting fraction is the numerator of the original fraction multiplied by $\frac{1}{m}$. □

3 Calculate and write each answer in its simplest form.

(a) $\frac{3}{4} \div 2 = $ □

(b) $\frac{8}{13} \div 4 = $ □

(c) $\frac{39}{56} \div 15 = $ □

(d) $\frac{21}{100} \div 84 = $ □

4 A team took 15 days to complete $\frac{3}{5}$ of a project. Answer the following questions.

(a) What fraction of the project did the team complete each day on average? □

(b) How many more days will it take to complete the whole project? □

Challenge and extension question

5 The figure below shows a rectangular strip of paper. If the whole strip represents 1, draw on the figure to show the meaning of $\frac{1}{2} \div 3$. Show your drawing and explain your reasoning.

6.11 Converting between fractions and decimals

Learning objective Convert between fractions and decimals

Basic questions

1 Multiple choice questions. (For each question, choose the correct answer and write the letter in the box.)

(a) Converting $2\frac{1}{10}$ to a decimal, the correct result is ☐.

    A. 0.2   **B.** 2   **C.** 2.01   **D.** 2.1

(b) Converting 1.25 to a fraction (including mixed number), the incorrect result is ☐.

    A. $1\frac{1}{4}$ **B.** $\frac{5}{4}$ **C.** $1\frac{25}{100}$ **D.** $1\frac{25}{1000}$

(c) In the following fractions, ☐ is closest to 0.43.

    A. $\frac{21}{50}$ **B.** $\frac{2}{5}$ **C.** $\frac{9}{20}$ **D.** $\frac{15}{25}$

2 (a) Convert each fraction to a decimal.

$\frac{1}{2} =$ _____ $\frac{3}{5} =$ _____ $\frac{7}{10} =$ _____

$1\frac{49}{100} =$ _____ $\frac{51}{50} =$ _____

(b) Convert each decimal to a fraction in its simplest form.

0.3 = ☐   0.02 = ☐   2.375 = ☐

5.16 = ☐   3.875 = ☐

3 Convert each fraction to a decimal by division or by using a decimal fraction (a decimal fraction is a fraction whose denominator is 10, 100, 1000, and so on). The first one has been done for you.

(a) $\frac{11}{20}$

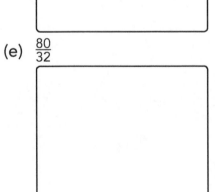

Solution 1:
Using division

$$20 \overline{)\,11.00} = 0.55$$

$$\frac{11}{20} = 0.55$$

Solution 2: Using decimal fraction
$$\frac{11}{20} = \frac{55}{100} = 0.55$$

(b) $\frac{5}{8}$

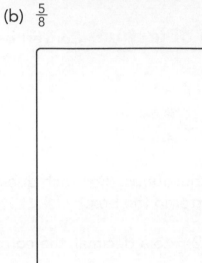

(c) $\frac{21}{40}$

(d) $\frac{13}{16}$

(e) $\frac{80}{32}$

(f) $\frac{48}{25}$

4 Convert the decimals to fractions, writing each fraction in its simplest form. The first one has been done for you.

(a) 0.28

(b) 0.05

(c) 0.36

$$0.28 = \frac{28}{100}$$
$$= \frac{7}{25}$$

(d) 8.75

(e) 5.432

(f) 7.84

5 A wheat factory produced 12.5 tonnes of flour on the first day. It produced $1\frac{1}{3}$ more tonnes of flour on the second day than on the first day. How many tonnes of flour did the plant produce in the two days?

6 Put the numbers 0.87, $\frac{4}{5}$, 0.88 and $\frac{7}{8}$ in order using <.

Challenge and extension question

7 As we learned in Chapter 2, a **recurring decimal** is a decimal number in which a digit or a sequence of digits in the decimal part repeats forever. For example, 0.111 111 … (the digit '1' repeats forever) and 0.313 131 … (the digits '31' repeat forever) are recurring decimals.

Convert the following fractions to recurring decimals and then round them to three decimal places. The first one has been done for you.

(a) $\frac{1}{3}$

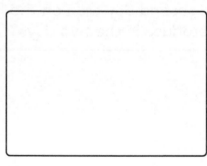

$$\frac{1}{3} = 1 \div 3$$
$$= 0.333\,333...$$
$$\approx 0.333$$

(b) $\frac{5}{9}$

(c) $\frac{15}{11}$

(d) $\frac{25}{99}$

6.12 Mixed operations with fractions and decimals

 Learning objective Solve mixed calculation problems with fractions and decimals

 Basic questions

1 Multiple choice questions. (For each question, choose the correct answer and write the letter in the box.)

(a) In one of the following operations, it is much easier to solve using fractions. Which one? ☐

    A. $\frac{2}{3} + 0.5$ **B.** $2.9 - \frac{2}{7}$ **C.** $3\frac{1}{8} - 1.2$ **D.** $0.31 + 2\frac{6}{11}$

(b) In one of the following operations, it would be particularly difficult to convert the fraction to an equivalent decimal. Which one? ☐

    A. $2\frac{3}{4} + 1.2$ **B.** $3\frac{1}{2} + 1.7$ **C.** $2.9 + \frac{2}{5}$ **D.** $3\frac{2}{3} - 2.5$

(c) In calculating $3\frac{7}{10} \times 1.5 + 8\frac{1}{2} \times 3.7$, it is easier to first ☐ .

    A. convert the decimals to fractions and then multiply

    B. convert the fractions to decimals and then multiply

    C. use the distributive law of multiplication over addition and then do the addition in the bracket

    D. use the associative law of multiplication

2 Calculate and write your answers in whole numbers or decimals.

(a) $1\frac{3}{5} - 0.7 =$ ☐   (b) $\frac{3}{4} \div 3 =$ ☐

(c) $5\frac{1}{8} - \left(3\frac{1}{8} - 0.375\right) =$ ☐

(d) $\frac{6}{7} + \frac{1}{7} \times \left(3.125 + 4\frac{7}{8}\right) =$ ☐

3 Calculate and write your answers to these calculations in fractions in their simplest form (write mixed numbers for improper fractions).

(a) $3\frac{3}{4} \times 1.6 + 5.5 \times \frac{5}{22}$

(b) $\frac{4}{3} \times \left(8\frac{7}{12} - 4\frac{1}{3}\right) \times \frac{5}{17}$

(c) $4.5 + \left(2\frac{1}{3} - 2.1 \times \frac{6}{7}\right)$

(d) $\left(2\frac{5}{6} + 1\frac{1}{3}\right) \div \left(2\frac{5}{6} - 1\frac{1}{3} + \frac{1}{2}\right)$

4 Find an easy way to calculate each of the following.

(a) $\left(\frac{1}{9} + 1\frac{4}{7} + \frac{8}{21}\right) \times 63$

(b) $\frac{5}{12} \times \frac{3}{7} + \frac{5}{12} \times \frac{4}{7}$

(c) $6.1 \times \frac{4}{5} - 2.6 \times \frac{4}{5}$

(d) $7.5 \times 5\frac{1}{2} + 7\frac{1}{2} \times 3.5 + 7\frac{1}{2}$

Challenge and extension question

5 Calculate the answers.

(a) $\frac{5}{13} + \left(\frac{6}{17} - \frac{3}{26}\right) \times \frac{2}{3} - \frac{4}{17}$

(b) $\left(16.25 - 13\frac{3}{4}\right) \times \frac{11}{12} + \frac{11}{12} \times 3\frac{1}{2}$

(c) $\left(15\frac{1}{5} - 2\frac{3}{10}\right) \times 77\frac{1}{6} + 12\frac{9}{10} \times 22\frac{5}{6}$

6.13 Application of fraction operations

Learning objective Solve mixed calculation problems with fractions

Basic questions

1 Complete each statement.

(a) If Number A is 5 and Number B is 4, then Number A is [ ] of Number B.

Number B is [ ] of Number A. (Fill in with a fraction.)

(b) The original price of an item was £180. It is now selling at $\frac{9}{10}$ of the original price. The current sale price is £[ ].

(c) The sale price of an item after a discount of $\frac{1}{10}$ is £180. The original price was £[ ].

(d) There are 444 pupils in a year group and 259 of them are boys. The number of girls is [ ] that of the boys. (Fill in with a fraction.)

(e) There were 420 pupils in a year group. The number of pupils was increased by $\frac{1}{6}$. There are now [ ] pupils.

(f) The number of pupils in a year group was increased by $\frac{1}{6}$ of the original number. If there are now 420 pupils, then there were [ ] pupils before.

(g) If the value of the output this month has increased by $\frac{1}{5}$, compared with the last month, then the value of the output last month was ⬜ that of this month. (Fill in with a fraction.)

(h) If Max's result in a 100-metre race was 15 seconds, and Ellie's result was $\frac{29}{30}$ of Max's, then Ellie was ⬜ seconds faster than Max in the 100-metre race.

(i) Ahuja has some money. After he spends $\frac{1}{6}$ of his money on a book and donates $\frac{9}{14}$ of the remainder to a charity, he has £25 left. Ahuja had £⬜ at first.

2 Bruce is reading a 200-page book. If he reads 11 more pages, then the number of pages he will have read is $\frac{3}{8}$ of the whole book. What fraction of the whole book has he read? He must return the book after 8 more days. How many pages should he read per day, on average, in 8 days so that he can finish the whole book before he returns it?

3 Jane is reading a book. If she reads 4 more pages, then the number of pages she will have read is $\frac{3}{10}$ of the whole book. If the book is due to be returned after 8 days, she must read 11 pages per day on average in the remaining 8 days to finish the book. How many pages are there in the book?

4 There are 720 tonnes of fruit. On the first day, $\frac{3}{8}$ of the fruit is delivered. How many tonnes of fruit should be delivered on the second day so that the remaining fruit is $\frac{1}{3}$ of all the fruit?

Challenge and extension questions

5 After $\frac{1}{3}$ of an empty bucket was filled up with oil, the total weight of the oil and the bucket itself was 12 kg. After it was fully filled up, the total weight became 30 kg. Find the weight of the bucket.

6 A lorry was used to deliver some goods. The whole journey for a round trip took 14 hours. The time taken to deliver the goods to the destination was $1\frac{1}{3}$ times the time taken to return. The average speed on the way to deliver the goods was 20 km per hour slower than that of the return. What was the distance the lorry travelled for the whole journey?

Chapter 6 test

1 Multiple choice questions. (For each question, choose the correct answer and write the letter in the box.)

(a) There are ☐ proper fractions in the simplest form with a denominator of 6.

    A. 1   **B.** 2   **C.** 3   **D.** 4

(b) Three pupils, A, B and C, were assigned to read the same article. It took Pupil A $\frac{6}{7}$ hours, Pupil B $\frac{7}{8}$ hours and Pupil C 0.8 hours to complete.

    The pupil who read the fastest was ☐.

    A. Pupil A   **B.** Pupil B   **C.** Pupil C   **D.** uncertain

(c) Mr Smith planned to process 50 spare parts in a day. He completed $\frac{3}{5}$ of his plan in the morning and $\frac{5}{7}$ of his plan in the afternoon. He has completed ☐ more of his plan.

    A. $1 - \frac{3}{5} - \frac{5}{7}$ **B.** $50 - \frac{3}{5} - \frac{5}{7}$ **C.** $1 + \frac{3}{5} - \frac{5}{7}$ **D.** $\frac{3}{5} + \frac{5}{7} - 1$

(d) The value of $7 \times \frac{1}{7} \div 7 \times \frac{1}{7}$ is equal to ☐.

    A. 1   **B.** $\frac{1}{49}$ **C.** 49   **D.** $\frac{1}{7}$

2 Complete each statement.

(a) $2012 \div 2013$ can be written in fraction form as ☐.

(b) In $\frac{2}{10}, \frac{12}{13}, \frac{3}{7}, \frac{21}{33}, \frac{15}{4}$ and $\frac{35}{42}$, the fractions in the simplest form

    are _____.

(c) Compare the numbers: $1\frac{4}{7}$ _____ $1\frac{3}{5}$ (Choose from >, < or =.)

(d) $4\,\text{m}^2\ 500\,\text{cm}^2 = $ ☐ m^2. (Fill in with a fraction.)

(e) The diagram below shows a number line. Point A represents the

number ⬚ and Point B represents the number ⬚ .

(Fill in each blank with a fraction.)

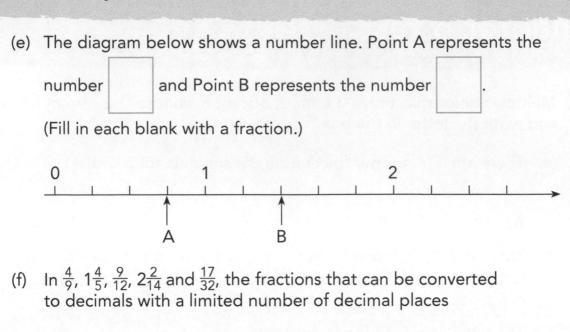

(f) In $\frac{4}{9}$, $1\frac{4}{5}$, $\frac{9}{12}$, $2\frac{2}{14}$ and $\frac{17}{32}$, the fractions that can be converted to decimals with a limited number of decimal places

are _____ .

(g) Joyce has read 80 pages of a book, which is $\frac{2}{5}$ of the whole book.

The book has ⬚ pages.

(h) There are two buckets. The water in Bucket A is $4\frac{2}{5}$ litres more than in Bucket B. If $3\frac{1}{3}$ litres of water from Bucket B is poured into Bucket

A, then there will be ⬚ litres more water in Bucket A than in Bucket B.

(i) The denominator of a fraction is 65. After simplification, it is $\frac{2}{5}$.

The fraction before simplification is ⬚ .

(j) Put $\frac{7}{8}$, $0.8\dot{7}$, $\frac{5}{6}$ and $\frac{15}{16}$ in order, from the least to the greatest:

_____ .

3 Calculate. Give your answers in whole numbers, mixed numbers or fractions in the simplest form.

(a) $2\frac{2}{3} \times 8 \times 1\frac{3}{8}$

(b) $3\frac{3}{5} - \frac{7}{13} - \left(0.6 + 1\frac{6}{13}\right)$

(c) $13.2 \times \frac{5}{6} \times \frac{15}{22}$

(d) $\frac{7}{40} \div 2 \times 6\frac{2}{3}$

(e) $4.35 \times 1\frac{2}{5} + 8\frac{3}{5} \times 4\frac{7}{20} + 1.5$

(f) $18\frac{1}{3} \div 5 \times \frac{9}{22} \div 3$

4 The diagram on the right shows a flowchart of operations. Answer the following questions.

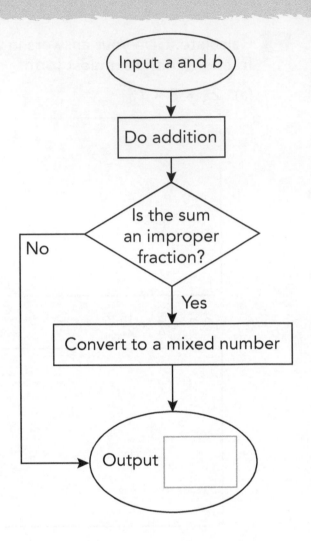

Input *a* and *b*

Do addition

Is the sum an improper fraction?

No

Yes

Convert to a mixed number

Output

(a) If the input numbers are $\frac{3}{5}$ and $\frac{4}{7}$, then what number is the output?

(b) If the input numbers are $\frac{2}{9}$ and $\frac{5}{18}$, then what number is the output?

5 A book has 240 pages. Sim read 6 pages more than $\frac{1}{6}$ of the whole book on the first day. On the second day, he read 8 pages fewer than $\frac{1}{8}$ of the whole book. How many pages has Sim not read yet?

6 There was a batch of tiles at a pool construction site. 7200 kg of the tiles were used in the first week, which was $\frac{3}{4}$ tonnes more than what was used in the second week. The remaining tiles were 0.8 tonnes more than the sum of the tiles used in the first two weeks. What was the weight of the batch of tiles? (Note: 1 tonne = 1000 kg.)

7 Jack and Lily went shopping for summer camp activities. They each had the same amount of money. Jack wanted to buy a camp bag and Lily wanted to buy a thermal bottle. After arriving in the store, they realised they didn't have enough money. If Jack gave Lily enough money to buy the bottle, he would have £20 left. If Lily gave Jack enough money to buy the camp bag, then she would have £8 left. The price of the bottle was $\frac{2}{3}$ that of the camp bag. Can you work out how much money Jack and Lily each had to start with? Show your working.

Chapter 7 Ratio, proportion and percentage

7.1 Meaning of ratio

 Learning objective Solve problems involving ratio

 Basic questions

1 Multiple choice questions. (For each question, choose the correct answer and write the letter in the box.)

(a) If there are 41 pupils in a class and 18 of them are girls, then the ratio of the number of boys to the number of girls is ☐ .

   A. $\frac{18}{41}$ **B.** $18:23$ **C.** $\frac{23}{18}$ **D.** $\frac{18}{23}$

(b) Surgical spirit, which is used in the practice of medicine, is a mixture of pure ethyl alcohol and water. If there are 75 litres of pure ethyl alcohol in every 100 litres of surgical spirit, the ratio of the volume of pure ethyl alcohol to the volume of water is ☐ .

   A. $1:3$ **B.** $3:1$ **C.** $1:4$ **D.** $4:1$

(c) If it takes Mark 15 minutes to cycle to school and 40 minutes to walk to school, then the ratio of Mark's average cycling speed to his average walking speed is ☐ .

   A. $1:40$ **B.** $40:1$ **C.** $3:8$ **D.** $8:3$

2 Complete each statement.

(a) In a ratio, the first quantity is 3 and the second quantity is 1. The ratio of the two quantities is _____ and the value of the ratio is _____ .

(b) The value of the ratio 20 cm : 1.2 m is _____.

(c) The value of the ratio $\frac{1}{5}$ hour : 12 minutes is _____.

(d) 'Pupil teacher ratio' refers to the ratio of the number of pupils to that of teachers. There are 420 pupils and 20 teachers in a school. The pupil teacher ratio in the school is _____.

3 The figure below shows two circles. The ratio of the radius of the smaller circle to that of the larger circle is 1 : 2. What is the ratio of the diameter of the smaller circle to that of the larger circle? _____.

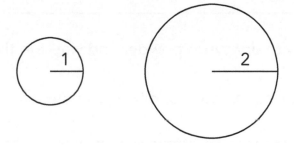

4 For each of the following pairs of similar shapes, what is the scale factor from the shape on the top to the shape on the bottom – that is, the ratio of the lengths of corresponding sides of the second (bottom) shape to those of the first (top) shape? Write your answers in the boxes. (Note: The pairs of figures are identical in shape and all the ratios of the lengths of their corresponding sides are equal.)

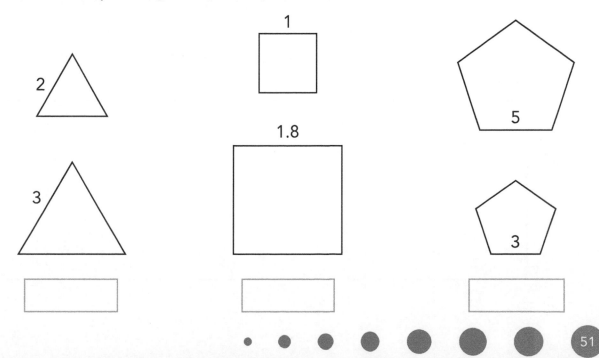

5 A landscaping team were planting a batch of trees. They planted $\frac{1}{5}$ of the trees on the first day and 136 trees on the second day. The ratio of the number of the remaining trees to the number of the trees that were planted was $3:5$. How many trees were there in the batch?

6 A medical syrup was formulated with dry syrup powder and water in the ratio of $3:400$.

(a) To make 1612 kg of the syrup, what amount of dry syrup powder is needed?

(b) If 60 kg of water is used, then what amount of dry syrup powder is needed?

(c) If 48 kg of dry syrup powder is used, then what amount of the medical syrup can be formulated?

Challenge and extension question

7 If the three angles in a triangle are in the ratio of $3:2:1$, then what type of triangle is it? Give your reason.

7.2 Properties of ratio

Learning objective Solve problems involving ratio

Basic questions

1 Multiple choice questions. (For each question, choose the correct answer and write the letter in the box.)

(a) Worker A took 12 days and Worker B took 18 days to complete the same job. The ratio of their work efficiencies in the simplest form is ⬜.

    A. 6:9   **B.** 3:2   **C.** 2:3   **D.** 9:6

(b) In the ratio 8:9, if the first number is increased by 16, in order to make the ratio unchanged, the second number must be ⬜.

    A. increased by 16   **B.** multiplied by 3

    C. unchanged   **D.** uncertain

(c) If $\frac{1}{2}x = \frac{1}{3}y = \frac{1}{4}z = $, then $x:y:z = $ ⬜.

    A. 2:3:4   **B.** 4:3:2   **C.** 6:4:3   **D.** 3:2:4

2 Complete each statement.

(a) Simplify the following ratios to the ratios of two integers in the simplest form.

> To simplify a ratio, we can apply the same method used to simplify a fraction until the two quantities of the ratio are integers with no common factors except 1.

    (i) $\frac{45}{63} = $ _____

    (ii) $\frac{1.35}{6} = $ _____

    (iii) $\frac{4}{9}:4 = $ _____

    (iv) 800 g : 2 kg = _____

    (v) 12 days : 2 weeks = _____

(b) Fill in the boxes with integers.

(i) $30:25 = \boxed{}:5$ (ii) $0.75:4.5 = 1:\boxed{}$

(iii) $81:\boxed{} = 9:5$ (iv) $76\,cm:57\,cm = \boxed{}:3$

(c) If $\frac{9}{10}$ of Number A is equal to 6 times Number B, then the ratio of

Number A to Number B in the simplest form is _____.

3 A box contains some balls coloured red, green and yellow. The number of red balls is $\frac{3}{4}$ of the number of green balls. The ratio of the number of green balls to that of yellow balls is $4:5$. The total number of green and yellow balls is 81. How many balls of each colour are there?

4 A rope was cut into two pieces in the ratio of $3:2$. The shorter piece was cut again into two pieces in the ratio of $2:1$. In all three pieces, what is the ratio of the length of the longest piece to that of the shortest piece?

5 To ice a cupcake, Worker A took 3 minutes, Worker B took 4 minutes and Worker C took $\frac{1}{2}$ minute. If the three workers iced a total of 372 cupcakes in the same period of time, then how many cupcakes did each worker ice?

 Challenge and extension question

6 Complete each statement.

(a) If $a:b:c = 2:3:5$ and $a = 20$, then $b = \boxed{}$ and $c = \boxed{}$.

(b) If $a:b = x:y$, $b:c = y:z$, then $a:b:c = $ _____.

Hence, if $a:b = 2:3$, $b:c = 3:7$, then $a:b:c = $ _____.

(c) $x:y = 0.8:1.2 = \boxed{}:6$, $y:z = 2:7 = 6:\boxed{}$,

so $x:y:z = $ _____.

(d) If $a:b = 2:3$, $b:c = 7:9$, then $a:b:c = $ _____.
Show your working.

7.3 Proportion

Learning objective Solve problems involving ratio

Basic questions

1 Multiple choice questions. (For each question, choose the correct answer and write the letter in the box.)

(a) In the following pairs of ratios, the two ratios that are not in proportion are ☐.

    A. $7 : 8$ and $14 : 16$ **B.** $0.6 : 0.2$ and $3 : 1$

    C. $0.9 : 1$ and $10 : 9$ **D.** $1 : 6$ and $\frac{2}{3} : 4$

(b) If $a : b = c : d$, then we have ☐.

    A. $a \times b = c \times d$

    B. $a \times c = b \times d$

    C. $a \times d = b \times c$

    D. none of the above

> The result we get from question (b) can be handy in solving problems about proportions. Can you describe it in words?

(c) Converting $2 \times 1.5 = 0.5 \times 6$ to a proportion, the correct one is ☐.

    A. $\frac{2}{1.5} = \frac{0.5}{6}$ **B.** $2 : 6 = 1.5 : 0.5$

    C. $6 : 1.5 = 2 : 0.5$ **D.** $6 : 0.5 = 1.5 : 2$

(d) The incorrect equation of proportion formed with four numbers 2, 5, 8 and 20 is ☐.

    A. $2 : 5 = 8 : 20$ **B.** $2 : 8 = 5 : 20$

    C. $5 : 8 = 2 : 20$ **D.** $5 : 2 = 20 : 8$

(e) The four numbers in ☐ cannot form a proportion.

A. 1, 2, 3, 6   B. 0.5, 1, 1.5, 3

C. $\frac{1}{2}$, $\frac{1}{3}$, 2, 3   D. 10, 20, 30, 40

2 Find x.

(a) $x : 13.5 = 2\frac{4}{5}$

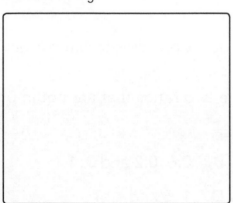

(b) $6\frac{1}{4} : x = 5$

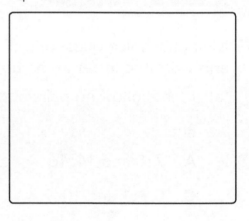

(c) $7\frac{1}{2} : 3 = x : 5$

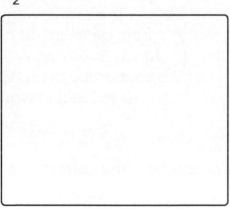

(d) $3 : \frac{1}{4} = \frac{3}{5} : x$

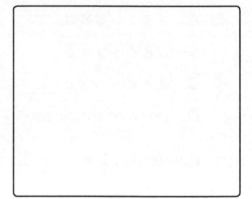

3 A box contains 100 balls with 40 red balls and 60 blue balls.

(a) What is the ratio of the number of red balls to the number of blue balls? Express your answer in its simplest form.

(b) David, Ivy and Mason each picked 20 balls from the box. David picked 9 red and 11 blue, Ivy picked 10 red and 10 blue, Mason picked 8 red and 12 blue. Which of them picked the balls proportionally to the ratio of the number of red balls and that of blue balls originally in the box? Give your reason.

4 Two branches of a company have different numbers of employees. If Branch A recruits 40 more employees, then the ratio of the number of employees in Branch A to that of Branch B is 2:1. If Branch B recruits 20 more employees, then the ratio of the number of staff in Branch B to that of Branch A is 5:6. How many employees does each branch have?

5 If the scale of a map is 1:1000000, and the measure of the distance between two cities on the map is 5.6 cm, then what is the actual distance between the two cities?

6 Jay made a scale drawing of his new house with a scale 1:50.

(a) In the drawing, the lounge is 8 cm × 12 cm. What is the actual size of the lounge? What is its area?

(b) The dining room is 4 m × 5 m. What is the measure of the dining room in the drawing?

Challenge and extension question

7 Find x. Show your working.

(a) $\dfrac{6.25}{x + 30} = \dfrac{1}{8}$

(b) $2 : 3 = (5 - x) : x$

7.4 Meaning of percentage

Learning objective Use equivalences between fractions and percentages to solve percentage problems

Basic questions

1 Multiple choice questions. (For each question, choose the correct answer and write the letter in the box.)

(a) If a whole is divided into 100 equal parts, then the incorrect statement is ☐ .

   A. One part is 1% of the whole.

   B. Two parts are 2% of the whole.

   C. Three parts are 3% of the whole.

   D. None of the above is correct.

(b) In a batch of products, the ratio of the number of non-defective products to that of defective products is 4 : 1. The percentage of defective products in this batch of products is ☐ .

   A. 25%   **B.** 20%

   C. 10%   **D.** none of the above

(c) In ☐ of the following four figures, the area of the shaded part in each figure is 50% of the area of the whole figure.

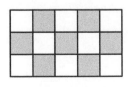 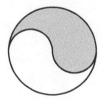 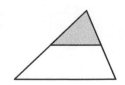

   A. one   **B.** two   **C.** three   **D.** all

(d) Given that item X weighs 60 kg and item Y weighs 54 kg, the correct

statement of the following is ☐ .

A. X is 10% heavier than Y.

B. Y is 11% lighter than X.

C. X is around 10% heavier than Y.

D. Y is 10% lighter than X.

2 Complete each statement.

(a) When converting a decimal to a percentage, we can move the decimal

point _____ places to the _____ and add the

percentage sign. When converting a percentage to a decimal,

we can move the decimal point _____ places to the

_____ and remove the percentage sign.

(b) If a shop offers a 25% discount on selected products, then the sale

prices are _____ % of the original prices.

(c) Convert the following numbers to percentages.

(i) $1.35 = $ _____   (ii) $\frac{7}{5} = $ _____

(iii) $2\frac{1}{4} = $ _____   (iv) $10\frac{7}{8} = $ _____

(d) Convert the following numbers to fractions in their simplest form.

(i) $45\% = $ _____   (ii) $3.2\% = $ _____

3 A farm harvested 400 000 kg of wheat the year before last. The production
of wheat last year was 20% more than the year before. Due to bad
weather conditions, the production this year was 20% less than last year.
What is the production of wheat this year?

_____

4 Convert each fraction first to a decimal and then to a percentage. Give your answer correct to three decimal places when rounding is needed.

(a) $\frac{2}{5}$ _____

(b) $\frac{7}{8}$ _____

(c) $\frac{5}{7}$ _____

(d) $\frac{11}{6}$ _____

5 Use fractions to express the quotients of the following divisions and then convert them to percentages. Give your answer correct to one decimal place before the percentage sign when rounding is needed.

(a) $52 \div 130$

(b) $28.4 \div 37$

6 If x is 25% of y and y is 80% of z, then how much percent is x of z?

7 If the length of one side of a square is decreased by 25% and the other side is increased by 2 m, then the area of the rectangle obtained is equal to the area of the original square. Find the area of the square.

7.5 Application of percentages (1)

Learning objective Solve percentage problems

Basic questions

1 Multiple choice questions. (For each question, choose the correct answer and write the letter in the box.)

(a) The greatest number of the following is ⬚.

    A. 37.5%   **B.** $\frac{1}{3}$ **C.** $\frac{7}{22}$ **D.** $\frac{\pi}{10}$

(b) In Jane's netball shooting practice, she scored 24 goals but missed 6 shots. The correct calculation to find the percentage of goals scored is ⬚.

    A. $\frac{6}{24} \times 100\%$ **B.** $\frac{6}{24+6} \times 100\%$

    C. $\frac{24}{24+6} \times 100\%$ **D.** $\frac{24}{6} \times 100\%$

(c) If the production of a factory increased by 20% last year compared with the previous year, and the production this year increased by 10% compared with last year, then the production this year increased by ⬚ compared with the year before last.

    A. 20% + 10%   **B.** 20% × 10%

    C. 20% − 10%   **D.** (1 + 20%) × (1 + 10%) − 100%

2 Complete each statement.

(a) A school library had 2500 books last year. The number of books was increased by 12% this year. The library now has ⬚ books.

(b) The output value of a factory this year increased by 20% compared with last year. If the output value of this year is £30 000 000, then the output value of last year was £⬚.

(c) The original price of a product was 30% less than its current price. If the current price is £560, then the original price was £[   ].

(d) If the price of a product after a 20% discount is £1200, then its original price was £[   ].

3 Year 6 had a maths test. One pupil was absent due to illness, 47 pupils passed (10 of them got distinction) and two pupils failed.

(a) What percentage of the class attended the test?

(b) What percentage of the pupils who took the test passed? (Give your answer correct to one decimal place before the percentage sign.)

(c) What percentage of the pupils who took the test got distinction? (Give your answer correct to one decimal place before the percentage sign.)

4 A school library had 3600 books last year and bought 504 new books this year.

(a) By what percentage is the number of books in the library increased this year compared with last year?

(b) What percentage of all the books now in the library are the new books bought this year? (Give your answer correct to one decimal place before the percentage sign.)

Challenge and extension question

5 In 2014, there were 34.4 million visits to the UK by overseas residents. In 2015 the number increased by 4% compared with 2014. From October 2015 to December 2015, the number was 8.47 million.
(Source: Office for National Statistics.)

(a) How many millions of visits were made by overseas residents to the UK in 2015? (Give your answer correct to two decimal places.)

(b) What percentage of the number of visits in 2015 is the number of visits from October 2015 to December 2015? (Give your answer correct to one decimal place before the percentage sign.)

(c) It was estimated that the number of visits by overseas residents to the UK would increase by 3.8% in 2016. Accordingly, how many millions of visits would be made by overseas residents to the UK in 2016? (Give your answer correct to two decimal places.)

7.6 Application of percentages (2)

Learning objective Solve percentage and ratio problems

Basic questions

1 Multiple choice questions. (For each question, choose the correct answer and write the letter in the box.)

(a) Delphi has read 60% of a book. The ratio of the number of pages she has not read to the number of pages of the book is ☐.

A. 2:3 **B.** 3:5 **C.** 2:5 **D.** 3:2

(b) A 6 km-long road is under repair. If 3.6 km of the road has been repaired, then ☐ of the road is unrepaired.

A. 60% **B.** 40% **C.** 36% **D.** 64%

(c) Jason was cycling on a road. If he had cycled 1.5 km and the remaining part was 25% of the whole distance, then he had ☐ more to cycle.

A. 6 km **B.** 0.5 km **C.** $\frac{3}{8}$ km **D.** $\frac{9}{8}$ km

2 Complete each statement.

(a) 30% of 180 is _____. 180 is 30% of _____.

(b) _____ is 20% more than 120. 120 is 20% more than _____.

(c) If Number A is $\frac{5}{7}$ of Number B, then Number B is ☐ of Number A.

(d) If Number A is 60% more than Number B, then Number B is _____% less than Number A.

3 The contents of a barrel of oil weighed 200 kg. If 7.5% of the oil was used on the first day and 10% of the remaining oil was used on the second day, how much oil was left in the barrel?

4 A school bought a number of books and distributed 60% of them to Years 4, 5 and 6 in the ratio of 2:3:4. Given that Year 6 was given 56 books, how many books did the school buy?

5 Air is a mixture of various gases, mainly oxygen and nitrogen. The other gases are carbon dioxide, water vapour and a number of rare gases. Nitrogen accounts for 78% of the volume of air, and the other gases including carbon dioxide, water vapour and a number of rare gases account for 1% of the volume of air. What percentage of the volume of air does oxygen account for?

Challenge and extension question

6 The figure shows the percentages of prize winners in four Year 6 maths classes in a maths competition in a school.

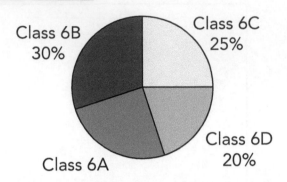

Class 6B 30%

Class 6C 25%

Class 6A

Class 6D 20%

(a) What percentage of all the prize winners are from Class 6A?

(b) By what percentage is the number of the prize winners in Class 6B more than Class 6D?

(c) Given that six prize winners are from Class 6B, how many prize winners are from Class 6C?

7.7 Application of percentages (3)

Learning objective Solve percentage problems involving the calculation of profit, loss and discount

Basic questions

1 Multiple choice questions. (For each question, choose the correct answer and write the letter in the box.)

(a) The original price of a jacket was £480. After a discount, the price was reduced by £120. The discount given was ☐.

A. 10% **B.** 20% **C.** 25% **D.** 30%

(b) If a store bought a product at a cost price of £400, set its regular retail price at £500, and finally sold it after a discount of 10%, then the percentage of profit from selling the product is ☐.

A. $\frac{500 - 400}{500} \times 0.9 \times 100\%$

B. $\frac{500 \times 0.9 - 400}{500} \times 100\%$

C. $\frac{500 - 400}{400} \times 0.9 \times 100\%$

D. $\frac{500 \times 0.9 - 400}{400} \times 100\%$

(c) The selling price of a computer desk is £50. If the percentage of profit is 40%, then the correct calculation to find the profit is ☐.

A. $50 \times 40\%$

B. $50 \div 40\% - 50$

C. $50 - 50 \div (1 + 40\%)$

D. $50 - 50 \div (1 - 40\%)$

2 Complete each statement.

(a) After an increase of 10% in the selling price of a TV set, the new price

is _____ % of the original price.

(b) If the original price of a set of golf clubs was £360 and the sale price

includes a discount of 25%, then the sale price is £_____
cheaper than the original price.

(c) If a set of hardback novels is sold at £120 after a discount of 20%, then

the original price of the set of novels was £☐.

(d) The selling price of a product was originally set 70% more than its cost price. For seasonal sales, it was put on sale for half price. According to the sale price, the percentage of loss in selling the product

is _____ %.

3 A department store bought a batch of clothes at a cost price of £200 per batch and the selling price was set to be 40% more than the cost price.

(a) What was the selling price of the clothes?

(b) During the sales period, the shop offered a discount of 20%. How much profit can the shop make in selling each batch of clothes?

4 A shop bought 300 diaries, intending to make a profit of 40% by selling each diary. After 60% of the diaries were sold, the shop made a profit of £540. What was the cost price per diary?

5 An electrical shop bought a batch of TVs. It would make a profit of £960 from selling each TV at the regular selling price. If the price is reduced by 20%, then it would make a loss of £832. What was the cost price per TV it bought?

6 One Christmas season, three furniture shops, A, B and C, launched various sales promotion schemes for a particular product with the same regular selling price of £4000.

Shop A: 'We offer a 13% discount.'

Shop B: 'You get £15 back for every £99 you spend in our store.'

Shop C: 'We offer a 10% discount, plus an extra £150 off.'

Based on these promotion schemes, answer the following questions.

(a) If Mr Johnson wants to buy the product, which shop should he go to? Why?

(b) If the original regular selling price, £4000, was 25% more than the cost price, could each shop still make a profit by selling the product during the promotion period? If so, which shop would have the greatest percentage of profit?

Chapter 7 test

1 Multiple choice questions. (For each question, choose the correct answer and write the letter in the box.)

(a) The four numbers in ☐ cannot form a proportion.

   A. 2, 3, 4, 6　　　　　　　　　　**B.** 1, 2, 3, 4

   C. 0.1, 0.3, 0.5, 1.5　　　　　　　**D.** $\frac{1}{2}, \frac{1}{3}, \frac{1}{4}, \frac{1}{6}$

(b) If Number A is 25% greater than Number B, then Number B is ☐ less than Number A.

   A. 25%　　　　**B.** 20%　　　　**C.** 22.5%　　　　**D.** 27.5%

(c) Sixty-four Year 6 pupils were divided into three groups, A, B and C in the ratio of $4:5:7$. After a new pupil joined Group B, the ratio of the number of pupils in Group B to that in Group C was ☐ .

   A. $3:4$　　　　**B.** $4:5$　　　　**C.** $5:6$　　　　**D.** $6:7$

2 Complete each statement.

(a) If $3a = 5b$ ($b \neq 0$), then $\frac{a}{b}$ = _____ .

(b) Simplify to the ratio of two integers in the simplest form.

   (i)　1.5 hours : 40 minutes = _____ .

(ii) $6.4:0.08 =$ _____ .

(c) Given $a:b = 2:3$, and $b:c = 4:5$, then $a:b:c =$ _____ .

(d) The measure of the distance between two places A and B on a map is 7 cm. If the scale of the map is $1:500\,000$, then the actual distance

between A and B is ⬚ km.

(e) A set of comic books is sold at a price of £240 after 20% off.

The original price of the set of comic books was £⬚ .

(f) The figure on the right shows the percentages of a family's expenses in a month. If the family's monthly expenses were £1000, then they spent

£ [ ] on education.

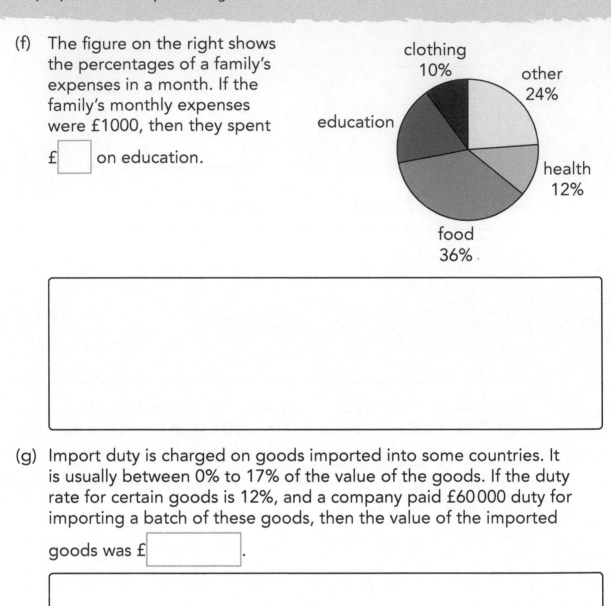

clothing
10%

other
24%

education

health
12%

food
36%

(g) Import duty is charged on goods imported into some countries. It is usually between 0% to 17% of the value of the goods. If the duty rate for certain goods is 12%, and a company paid £60 000 duty for importing a batch of these goods, then the value of the imported

goods was £ [ ].

3 Given $5x : 3 = 1\frac{1}{4} : 5$, find the value of x.

4 Given $x : y = 2 : 3$ and $y : z = 6 : 8$, find $x : y : z$.

5 The original price of a European tour package with a travel agency was £2800. The agency then offered it with a discount of 15%. How much cheaper was the discounted price than the original price?

6 A shop bought a batch of shoes at a cost price of £250 per pair. The selling price was 40% more than the cost price.

(a) What is the selling price for each pair of shoes?

(b) The shop then offered a discount of 20% off the selling price. How much profit could the shop still make by selling each pair of shoes?

7 The scale of a map is 1 : 3 000 000, and the measurement of the distance between two places, A and B, is 4.5 cm. A car starts travelling from A to B at 10 o'clock in the morning at a speed of 60 km per hour. Can the car reach B before 12:00 noon? Give your reason.

8 A government announced a policy to help first-time house buyers. Housing tax was 2% of the purchase price of the house. The policy offered a refund of 60% of the housing tax. If a first-time buyer bought a flat for £180 000 before the tax, how much did the buyer pay after the refund?

9 A shop sells some chairs at £60 per chair. The shop offered the following discount plan during a sale.

Total price of purchase (before discount)	Discount offered
£300 or less	no discount
more than £300 and up to £400	20% off
more than £400 and up to £500	30% off
more than £500 and up to £600	40% off

(a) Mr Wilton paid £360 for some chairs during the sale. How many chairs did Mr Wilton buy?

(b) Imagine you are the owner of the shop. Design a promotion plan so that Mr Wilton can buy 15 chairs for £450. Explain your plan.

Chapter 8 Geometry and measurement (II)

8.1 Nets of 3-D shapes (1)

 Learning objective Describe and build nets of cubes and cuboids

 Basic questions

1 Complete each statement.

(a) The net of a 3-D shape shows what the 3-D shape would look like if it were opened out flat. A net is a _____ shape. (Choose: 2-D or 3-D.)

(b) A cube is a 3-D figure made up of _____ identical squares. A net of a cube consists of six identical _____.

(c) A cuboid has _____ faces, which are usually rectangles. It is also possible for a cuboid to have _____ square faces. A cuboid can have different nets, but they all have two pairs of identical _____.

(d) Six identical non-square rectangles _____ be folded into a cuboid. (Choose: 'can' or 'cannot'.)

(e) A 3-D shape _____ have more than one net. (Choose: 'can' or 'cannot'.)

2 Which of the following figures can be folded to form a cube? Write the number of each figure. _____

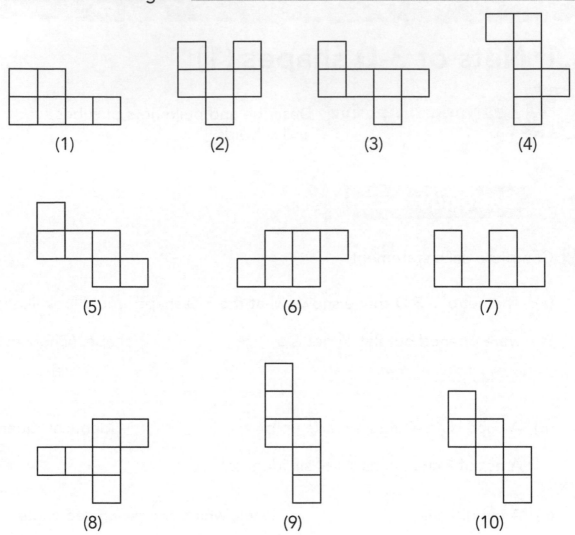

(1)   (2)   (3)   (4)

(5)   (6)   (7)

(8)   (9)   (10)

3 The figure shows the net of a cube. If it is folded back into the original cube, which points are adjacent to Point *P*?

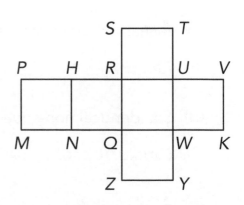

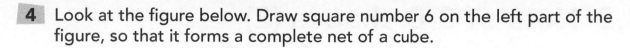

4 Look at the figure below. Draw square number 6 on the left part of the figure, so that it forms a complete net of a cube.

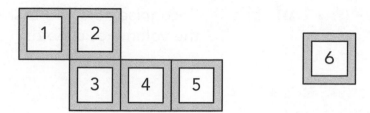

5 Multiple choice question. (Choose the correct answer and write the letter in the box.)

The first figure below shows a cubic paper box with a triangle, a square and a circle on three of its faces. If the box is cut along its edges and unfolded to form a net, it would be figure ☐ .

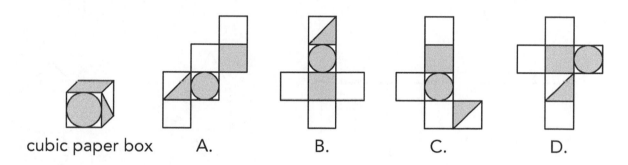

cubic paper box   A.   B.   C.   D.

Challenge and extension question

6 A lizard at vertex A of a cube wants to catch a mosquito at vertex B along the surface of the cube. Which is the shortest path for it to take?

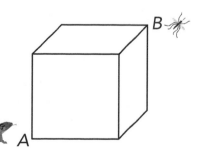

8.2 Nets of 3-D shapes (2)

Learning objective Recognise nets of 3-D shapes and calculate the volume of cuboids

Basic questions

1 Draw a line to join each 3-D shape with its net.

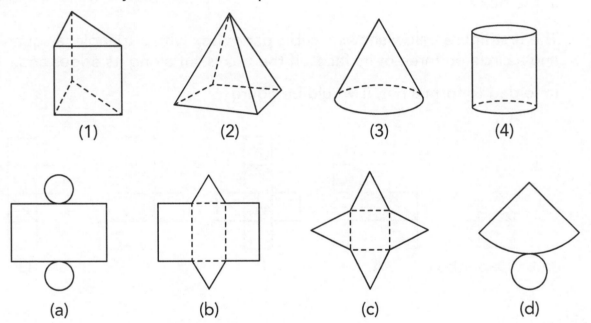

(1)   (2)   (3)   (4)

(a)   (b)   (c)   (d)

2 Circle the figures which can be folded to form a cuboid.

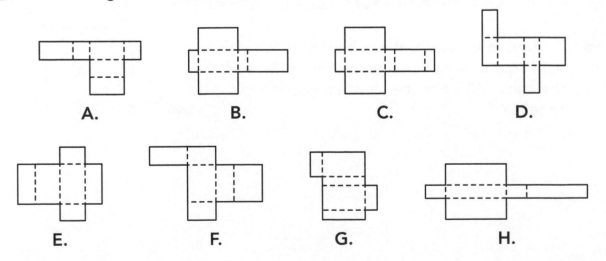

A.   B.   C.   D.

E.   F.   G.   H.

3 The figure below shows a net of a cuboid paper box. The length of the cuboid is 9 cm. Find the volume of the cuboid paper box.

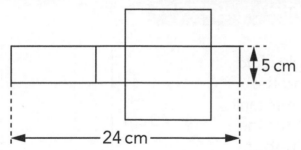

5 cm

24 cm

4 Each of the figures shows a net of a cuboid. Find the volume of each cuboid.

(a)

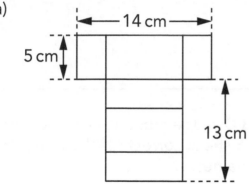

14 cm

5 cm

13 cm

(b)

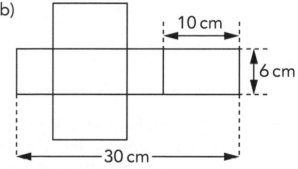

10 cm

6 cm

30 cm

Challenge and extension questions

5 As shown in the figure on the right, fold a square piece of iron sheet along the dotted lines to form a cuboid water tank with an extra piece of iron sheet for the bottom. What is the area of the bottom piece? How many cubic centimetres of water can be stored in the water tank?

40 cm

6 Sonia said to her sister Jane: 'Every 3-D shape must have a net, and you just need to open it flat.' Do you think she is correct? If she is incorrect, name a 3-D shape that does not have a net.

8.3 Reflection of shapes

 Learning objective Solve symmetry, reflection and angle problems

 Basic questions

1 Complete each statement.

(a) Look at the following figures.
Identify each figure in which you can use a line to divide it into two parts so when it is folded along the line, the two parts coincide with each other.

(Note: To 'coincide' means to occupy exactly the same place, or lie exactly on top of each other, in this context.)

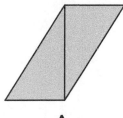 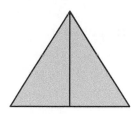 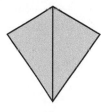 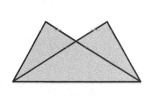

A.　　　　　　　　B.　　　　　　　　C.　　　　　　　　D.

Answer: _____.

(b) In Figure B above, the two parts divided by the line are

_____ and one part can be considered as a

_____ of the other part over the line. The figure is called a

_____ figure and the line is called a line

of _____.

(c) Look at each pair of figures below. In which pair can one figure be considered as a reflection of the other figure over a line? Write the letter in the box. ☐

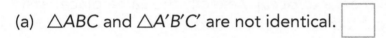

A.　　　　　**B.**　　　　　**C.**　　　　　**D.**

---

2 True or false? (Put a ✓ for true and a ✗ for false in each box.)

In the diagram, △A′B′C′ is the image of △ABC after a reflection in line *l*.

(a) △ABC and △A′B′C′ are not identical. ☐

(b) △ABC and △A′B′C′ are symmetrical with respect to the line *l*. ☐

(c) If A′C′ = 10, then AC = 10. ☐

(d) If ∠B = 80°, then ∠B′ = 80°. ☐

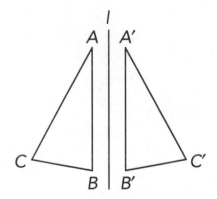

---

3 The diagram shows a 2 × 2 square grid and △ABC with A, B and C being nodes of the grid. Draw the following:

(a) The reflection image of △ABC in line CD.

(b) The reflection image of △ABC in line EF.

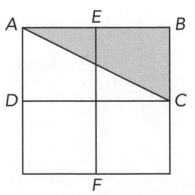

4 Look at the following figures carefully. Based on the pattern, draw a suitable figure in the blank space.

   _____

Challenge and extension questions

5 Jim is standing in a room, facing a mirror on a wall. Behind him, there is a digital clock on the wall opposite the mirror. He sees the reflection of the digital clock, which is as follows:

What is the actual time the digital clock shows? _____

6 In the figure on the right, ∠2 = 60°. In order to make the white ball first strike the cushion and then, after a reflection, strike the black ball and pot it to the pocket on the corner,

∠1 must be _____ .

A. 30°   B. 45°

C. 60°   D. 70°

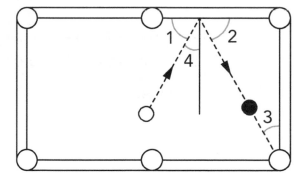

8.4 Translation of shapes

Learning objective Translate shapes and describe their movement accurately

Basic questions

1 Multiple choice questions. (For each question, choose the correct answer and write the letter in the box.)

(a) In the following figures, the one that can be viewed as the result of a translation is ☐.

A.

B.

C.

D.

(b) The square ABCD has a side length of 7 units. It is translated by 4 units in the direction of BC to give the square EFGH. The length of AH is ☐ units.

A. 9   **B.** 10   **C.** 11   **D.** 12

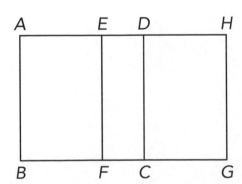

(c) As shown in the diagram, △A′B′C′ is obtained by a translation of △ABC.

The incorrect statement below is ☐.

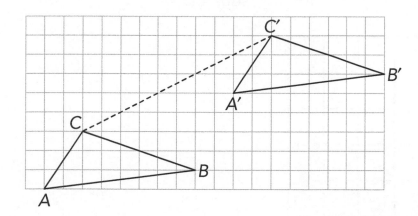

A. Translating △ABC 10 units right and then 5 units up gives △A′B′C′.

B. Translating △ABC 5 units up and then 10 units right gives △A′B′C′.

C. Translating △ABC in the direction of CC′ by the distance being the length of the line segment CC′ gives △A′B′C′.

D. Translating △ABC in the direction of C′C by the distance being the length of the line segment C′C gives △A′B′C′.

2 Complete each statement.

(a) If △A′B′C′ is the image of △ABC after a translation, and the images of points A, B and C are A′, B′ and C′, then A′B′ = _____ ,

∠A′ = _____ .

(b) In the diagram below, $O_1A = AB = BO_2 = 2\,cm$.

After the circle $\odot O_1$ is translated right ☐ cm, it coincides with $\odot O_2$.

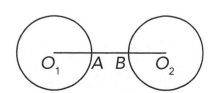

(c) The diagram below shows a screen of a tile-matching game. To translate the tile in the upper-left corner to the blank area in the lower part, it should move _____ for ☐ square(s), and then move _____ for ☐ squares.

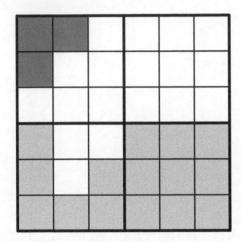

3 The diagram shows a square grid with △ABC. Translate △ABC so that point A moves to point A'. Indicate the images of points B and C, label them B' and C' and draw the translation image.

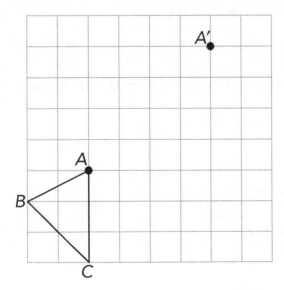

4 The diagram shows a square grid and a part of a circle with centre O. Translate the part of the circle to the position so that point O coincides with point A. Draw the translation image.

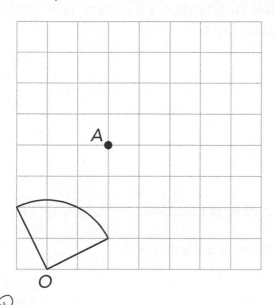

A●

O

Challenge and extension question

5 The diagram shows some rectangular land with a length of 12 m and a width of 8 m. If it is used to build a public green space (shaded region) with small paths measuring 2 m wide, what is the area of the green space, that is, the shaded region? Show your working.

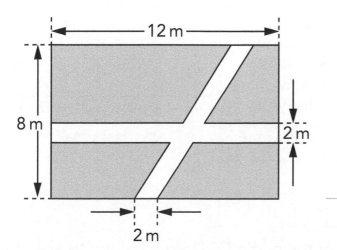

8.5 Coordinates and quadrants

Learning objective Describe positions on a coordinate grid in four quadrants

Basic questions

1 Complete each statement.

(a) On a coordinate grid, there are _____ quadrants: Quadrant I, Quadrant II, Quadrant III and Quadrant IV; there are also

_____ axes: a horizontal axis called the _____

and a vertical axis called the _____ .

(b) If $O(x, y)$ is the origin, then $x = $ _____ and

$y = $ _____ .

(c) If point $P(x, y)$ is on the x-axis, then $y = $ _____ .

(d) If point $P(x, y)$ is on the y-axis, then $x = $ _____ .

2 Look at the coordinate grid on the right.

(a) Use coordinates to describe the positions of points A, B, C, D, E and O. Which quadrant does each point belong to? Fill in the table below. The first one has been done for you.

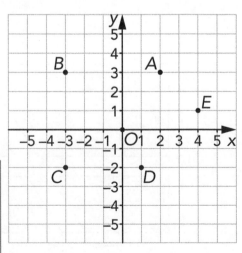

	A	B	C	D	E
Coordinates	(2, 3)				
Quadrant	I				

(b) What are the coordinates of the origin O? Does it belong to any

quadrant? _____

3 Mark the following points on the coordinate grid below.

M(1, 5), N(–2, –2), P(0, –3), Q(6, –6), R(–3, 0)

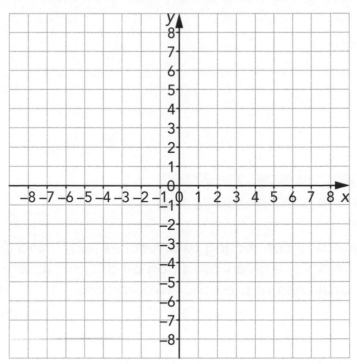

4 Multiple choice questions. For each question, choose the correct answer and write the letter in the box. (Hint: you may draw a diagram to help you.)

(a) If point P(a, b) is in the first quadrant, then point Q(–a, –b) is in the [ ].

 A. first quadrant

 B. second quadrant

 C. third quadrant

 D. fourth quadrant

(b) If point M(a, b) is in the second quadrant, then N(–a, –b) is in the [ ].

 A. first quadrant

 B. second quadrant

 C. third quadrant

 D. fourth quadrant

(c) If $E(a, b)$ is in the fourth quadrant,

then $F(b, a)$ is in the ☐.

   A. first quadrant

   B. second quadrant

   C. third quadrant

   D. fourth quadrant

Challenge and extension question

5 Given that a square has one side connecting two vertices $(-2, 0)$ and $(2, 0)$, draw all the squares that satisfy the given condition on the coordinate grid below. Write down the coordinates of the other two vertices.

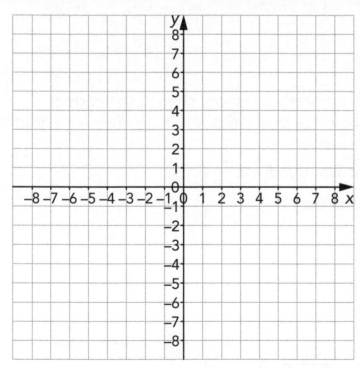

8.6 Reflection on the coordinate plane

Learning objective Reflect shapes on the coordinate plane

Basic questions

1 Given that a square *ABCD* has four vertices, *A*(0, 0), *B*(2, 0), *C*(2, 2) and *D*(0, 2), draw and label the square on the coordinate plane on the right.

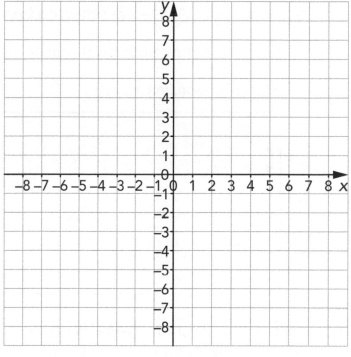

2 Given that a parallelogram has three known vertices *A*(−4, −4), *B*(6, 1) and *C*(−6, 2), and the fourth vertex *D* is in the first quadrant, draw and label the parallelogram on the coordinate plane on the right. What are the coordinates of point *D*?

_____

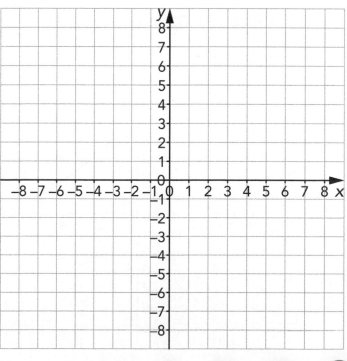

3 Complete the statements. (Hint: draw a diagram to help.)

(a) After point $A(3, -5)$ is reflected in the x-axis, its image is in the

_____ quadrant and the coordinates are _____.

(b) After point $A(3, -5)$ is reflected in the y-axis, its image is in the

_____ quadrant and the coordinates are _____.

(c) After point $A(3, -5)$ is reflected first in the x-axis and then in the

y-axis, the new position it reaches is in the _____ quadrant

and the coordinates are _____.

4 Look at the diagram. $\triangle ABC$ is on the coordinate plane. Draw and label the images of the following reflections.

(a) Reflect $\triangle ABC$ in the x-axis.

(b) Reflect $\triangle ABC$ in the y-axis.

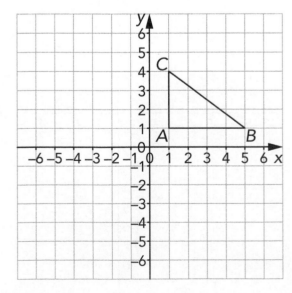

Challenge and extension question

5 Given that a parallelogram has three known vertices $A(0, 0)$, $B(-2, 7)$ and $C(-6, 7)$ on the coordinate plane below, answer the following questions.

(a) What could be the coordinates of its remaining vertex D? Write the coordinates of D. (Hint: think of all the possible locations of D.)

_____

(b) Draw the parallelogram(s) on the coordinate plane.

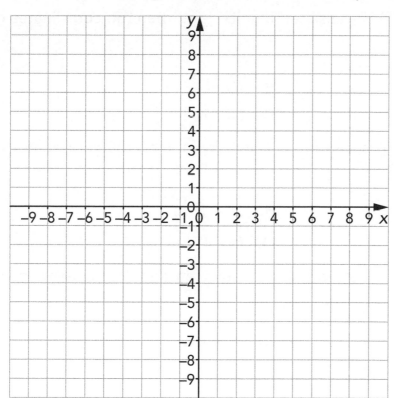

8.7 Translation on the coordinate plane

Learning objective Translate shapes on the coordinate plane

Basic questions

1 Multiple choice questions. For each question, choose the correct answer and write the letter in the box. (Hint: You may draw a coordinate plane to help you.)

(a) If point $P(2, 1)$ is translated 5 units right, then the coordinates of its image P' are ☐ .

    A. $(2, 6)$ **B.** $(7, 1)$

    C. $(2, -4)$ **D.** $(-3, 1)$

(b) If point $P(a, b)$ is translated 1 unit down, then the coordinates of its image P' are ☐ .

    A. $(a, b + 1)$ **B.** $(a + 1, b)$

    C. $(a, b - 1)$ **D.** $(a - 1, b)$

(c) Point $A(x, y)$ is first translated 2 units left and then 3 units down. The new position of the point is at ☐ .

    A. $(x + 2, y + 3)$ **B.** $(x + 2, y - 3)$

    C. $(x - 2, y + 3)$ **D.** $(x - 2, y - 3)$

(d) Point $A(2a - 4, a + 5)$ is translated 2 units down to reach point B. If B is on the x-axis, then the coordinates of A are ☐ .

    A. $(2, 8)$ **B.** $(0, 8)$

    C. $(-10, 2)$ **D.** $(-10, 0)$

2 Complete each statement. (Hint: You may draw a coordinate plane to help.)

(a) Translate point $A(2, 3)$ by 3 units right to get point B. The coordinates of B are _____.

(b) Translate point $A(2, 3)$ by 3 units down to get point B. The coordinates of B are _____.

(c) Translate point $A(2, 3)$ _____ units _____ to get point B. The coordinates of B are $(2, -1)$.

(d) Translate $A(-5, -6)$ by 6 units right and then 7 units up to get point B. The coordinates of B are _____.

(e) Translating $A(2, 3)$ first _____ unit(s) _____ and then _____ unit(s) _____, it reaches point $B(-3, 2)$.

(f) In order to translate $A(-2, 3)$ to the origin, we can first translate it _____ unit(s) _____ and then _____ unit(s) _____.

3 After point $A(a, a + 5)$ is translated 6 units right, it reaches point B. If B is also the reflection image of point A in the y-axis, then what are the coordinates of point A? Show your working. (Hint: You may draw a coordinate plane to help.)

_____

4 Parallelogram $ABCD$ is on the coordinate plane as shown. Draw and label the images of the following translations, respectively.

(a) Translate the parallelogram 6 units left.

(b) Then translate the parallelogram 10 units down.

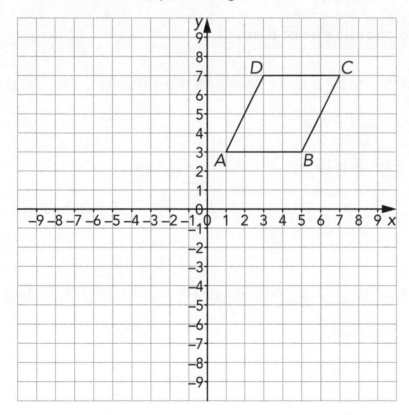

5 On the coordinate plane below, draw a circle with centre A(–3, 2) with radius of 4 units and then translate it 6 units right. Draw its image of the translation.

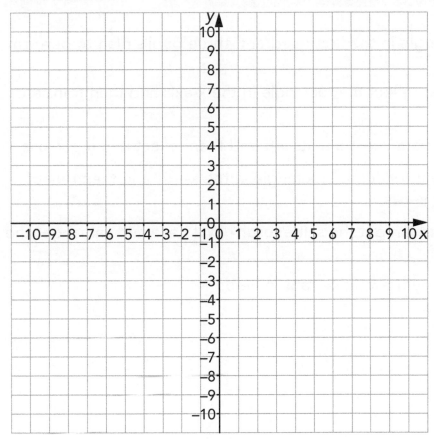

Challenge and extension question

6 Given any rectangle in the first quadrant, three pupils plan to perform the following.

(a) Kaya will first reflect the rectangle in the x-axis and then translate it 5 units left.

(b) Ian will first translate it 5 units left and then reflect it in the x-axis

(c) Jamal will first reflect the rectangle in the x-axis and then reflect it again in the y-axis.

Will Kaya and Ian get the same result? Will Kaya and Jamal get the same result?

(Hint: you may draw a diagram to help.)

_____

_____

1 Complete each statement.

(a) A net of a cube consists of six _____, and a net of a cuboid

consists of six _____.

(b) A net of a square-based pyramid consists of _____ square(s)

and _____ triangles.

(c) When a shape is reflected or translated to a new position, the shape

_____ changed. (Choose: 'has' or 'has not'.)

(d) In terms of geometric property, the figure below that is different from

the other three is figure number ☐.

 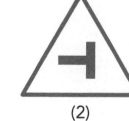 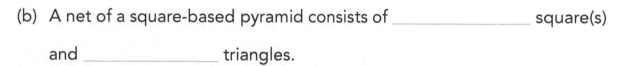 

(1)   (2)   (3)   (4)

(e) The diagram below shows two triangles. After △ABC is translated to

the right, it coincides with △DFE. If $BF = 4$, then ☐ = 4.

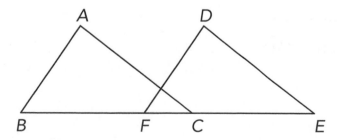

(f) Translating point $P(3, -2)$ first 2 units down and then 6 units left to get

point B, the coordinates of B are _____.

(g) Translating point $P(3, -2)$ first ☐ unit(s) to the _____ and then ☐ unit(s) up, it reaches point $(-8, 10)$.

(h) Translating point $P(x, y)$ a units down to get point B, the coordinates of B are _____.

(i) Reflecting point $P(x, y)$ in the x-axis to get point Q and then translating Q a units left to get point M. The coordinates of M are _____.

2 Multiple choice questions. (For each question, choose the correct answer and write the letter in the box.)

(a) Point $A(-3, 1)$ is first translated 4 units left and then 5 units down to get point B. The coordinates of point B are ☐.

   A. $(1, 6)$

   B. $(1, -4)$

   C. $(-7, 6)$

   D. $(-7, -4)$

(b) If point $A(a, b)$ is in the fourth quadrant, then $B(-a, b)$ is in the ☐.

   A. first quadrant

   B. second quadrant

   C. third quadrant

   D. fourth quadrant

(c) Each of the following figures is divided by line *l* into two parts. The figure in which one part can be viewed as the image of the other part by a reflection in line *l* is ☐ .

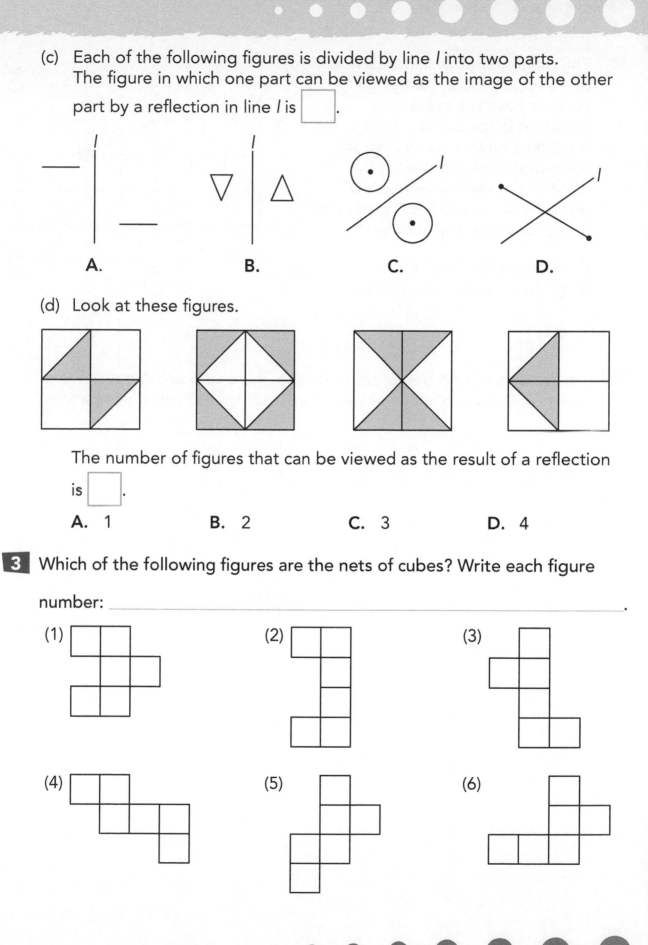

A.   B.   C.   D.

(d) Look at these figures.

The number of figures that can be viewed as the result of a reflection is ☐ .

A. 1   B. 2   C. 3   D. 4

3 Which of the following figures are the nets of cubes? Write each figure number: _____ .

(1)   (2)   (3)

(4)   (5)   (6)

4 The diagram below is the surface of a snooker table. All the small squares have the same side length. A black ball is placed in the position shown. After it is struck by a white ball, it moves in the direction of the arrow. In which pocket will the black ball land after reflections over the cushions?

Draw the reflections on the diagram to show your answer.

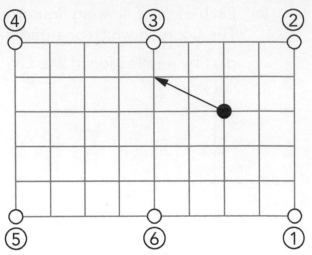

5 On the square grid below, translate the figure so point A translates to point A'. Draw the image of the figure after the indicated translation.

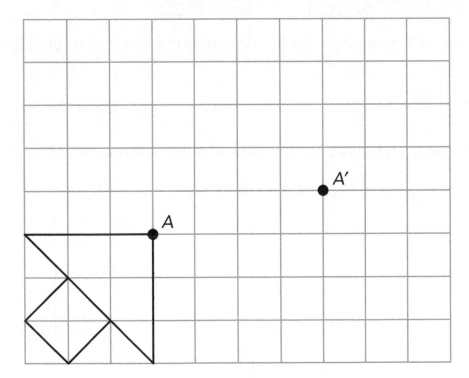

6 On the coordinate plane on the right, reflect two points A(6, 4) and B(–4, 4) in the x-axis to get points D and C respectively.

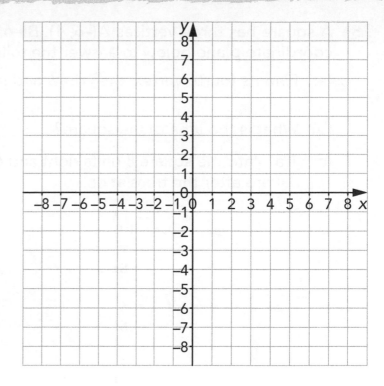

(a) Draw and label the four points as indicated and connect them to get quadrilateral ABCD.

(b) What shape is quadrilateral ABCD?

_____

(c) Find the area of quadrilateral ABCD.

_____

7 The diagram below shows △ABC on a 1 cm square grid (drawing not to scale).

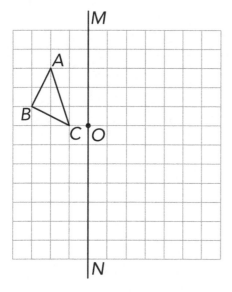

(a) Draw △A₁B₁C₁ that is the image of △ABC after a reflection in line MN.

(b) Draw △A₂B₂C₂ that is the image of △A₁B₁C₁ after a translation of 5 cm down and 2 cm right.

(c) Can △A₂B₂C₂ be obtained by directly translating △ABC?

_____

8 A square has three vertices $A(-6, 1)$, $B(-6, -3)$ and $D(-2, 1)$. Use the coordinate plane below to answer the following.

(a) Find its fourth vertex, then draw and label the square.

(b) What is the area of the square? _____

(c) Translate the square 8 units right and 6 units up. Draw the image of the square after the translation.

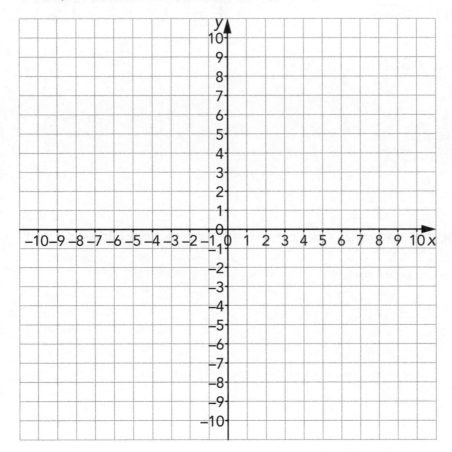

Chapter 9 Statistics (IV)

9.1 Mean

Learning objective Calculate and interpret the mean as an average

Basic questions

1 Complete each statement.

(a) Mean = the _____ of all the items ÷ the _____

of all the items. Mean is often also called _____.

(b) There are five packs of sugar with a total weight of 560 g. The mean

weight per pack of sugar is ⬚ g.

(c) A farm has 150 pigs. They weigh 9000 kg in total. The mean weight of

each pig is ⬚ kg.

(d) In mid-term tests, Joan scored 99 marks in Maths, 96 marks in English
and 90 marks in Science. Joan's mean score in these three subjects is

⬚ marks.

2 Find the mean of the numbers or quantities in each group.

(a) 56, 90, 108, 312

(b) 72 kg, 94 kg, 159 kg, 511 kg, 788 kg

3 The table shows the amount of money Joe's family spent on gas and electricity bills in a year.

Quarter	first	second	third	fourth
Amount (£)	566	308	429	1027

(a) How much did Joe's family pay in total for gas and electricity bills in the whole year?

(b) On average, how much did Joe's family pay per quarter?

4 The table shows the number of Year 6 pupils in school joining different extra-curricular clubs.

Club	football	basketball	table tennis	netball	film
Number of pupils	8	6	10	12	4

On average, how many Year 6 pupils joined each club?

5 The table shows the height and weight of six pupils in a class.

Pupil	Alvin	Barry	Lucy	Neo	Mandy	Jack
Height (cm)	138	152	140	145	155	143
Weight (kg)	33	41	38	44	43	35

(a) What is the mean height of these six pupils?

(b) What is their mean weight?

6 John scored the following marks in six maths tests last year: 98, 94, 92, 95, 100 and 97. What was his mean score for the six tests?

Challenge and extension question

7 There were 15 winners in a Maths competition and the marks they scored were 98, 90, 88, 84, 90, 92, 95, 89, 94, 91, 90, 92, 88, 81 and 91, respectively. What was the mean score of the winners?

9.2 Calculating the mean (1)

 Learning objective Solve problems involving finding the mean as an average

 Basic questions

1 Complete each statement.

(a) Four identical glasses are filled with water. The heights of the water levels are 3.2 cm, 4.5 cm, 5.5 cm and 6.4 cm. The mean height of the water levels in the glasses is ⬚ cm.

(b) A school choir has five groups. The numbers of pupils in the five groups are 37, 40, 39, 36 and 33. The mean number of pupils in each group is ⬚.

(c) The distances Tom ran in 5 days were 650 m, 880 m, 1050 m, 0 m and 930 m. The mean distance he ran per day in these 5 days was ⬚ m.

2 The table shows the amount of milk Lynn drank in a week.

Day	Monday	Tuesday	Wednesday	Thursday	Friday	Saturday	Sunday
Milk (ml)	150	200	0	180	0	220	160

How much milk did Lynn drink per day on average in the week?

3 A company delivered 20 tonnes of goods in the morning and 26 tonnes in the afternoon on the first day, 47 tonnes on the second day and 39 tonnes on the third day. On average, how many tonnes of goods were delivered per day in these three days?

4 The numbers of toys made by a factory in a week (5 working days) were 732, 698, 631, 0 and 1254. On average, how many toys were made per day in the week?

5 A car travelled from Place A to Place B. It travelled 160 km in the first 2 hours and 210 km in the next 3 hours. On average, how many kilometres did the car travel in 1 hour?

6 A grocery store received a batch of fruit. It sold 150 kg in the first 3 days, 160 kg in the next 5 days, and 350 kg in the last 7 days. How many kilograms of fruit did the store sell per day on average?

7 The table shows the number of times Leo took part in community activities in the first half of a year.

Month	January	February	March	April	May	June
Participation (number of times)	2	3	2	0	2	1

How many times did Leo participate in community activities each month on average? (Keep your answer to one decimal place.)

8 A delivery team delivered rice to a supermarket. On the first day, the team made 7 trips and delivered 1550 kg of rice. On the second day, it made 8 trips and delivered 1750 kg of rice. How many kilograms of rice did the team deliver per day on average in the two days? How many kilograms of rice did the team deliver per trip on average?

Challenge and extension question

9 There are six numbers with a mean of 48. After removing one of the numbers, 56, what is the mean of the remaining numbers?

9.3 Calculating the mean (2)

Learning objective Solve problems involving finding the mean as an average

Basic questions

1 Multiple choice questions. (For each question, choose the correct answer and write the letter or letters in the box.)

(a) There are 6 boxes of table tennis balls in a sports room. The number of table tennis balls that each box contains is 4, 6, 6, 6, 7 and 7, respectively. How many table tennis balls does each box contain on average? The correct expression is [ ]. (Choose all the correct ones.)

    A. $(4 + 6 + 6 + 6 + 7 + 7) \div 6$

    B. $(4 + 6 \times 3 + 7 \times 2) \div 3$

    C. $(4 + 6 + 7) \div 3$

    D. $(4 + 6 \times 3 + 7 \times 2) \div 6$

(b) Class A and Class B in Year 6 made paper flowers during an activity. In 4 hours, Class A made 120 paper flowers while Class B made 128 paper flowers. On average, how many flowers did each class make per hour? The correct expression is [ ].

    A. $(120 + 128) \div 2$

    B. $(120 + 128) \div 4$

    C. $(120 + 128) \div 4 \div 2$

    D. $(120 \times 4 + 128 \times 4) \div 2$

2 It took Joan 7 days to finish reading a graphic novel. The numbers of pages she read in these seven days were 24, 24, 22, 24, 21, 22 and 24, respectively. How many pages did she read each day on average?

3 A car travelled from Place A to Place B. In the first 2 hours, it travelled 65 km per hour, and in the next 3 hours, it travelled 75 km per hour. How many kilometres did the car travel per hour on average in the whole journey?

4 There are 20 boys and 16 girls in a Year 6 class. In an English test, the boys' mean score is 92 marks and the girls' is 88 marks. What is the mean score of the whole class in the test? (Round your answer to the nearest tenth.)

5 The table shows the numbers of skips Tom completed during 10 skipping rope exercises.

Time	1st	2nd	3rd	4th	5th	6th	7th	8th	9th	10th
Number of skips	35	35	35	48	50	35	50	34	48	50

How many skips did Tom do each time on average?

6 The table shows the numbers of boys and girls, as well as their mean weights, in two classes. Answer the following questions. (Round your answers to the nearest tenth.)

Boys' and girls' weights in Class A and Class B of Year 6

	Gender	No. of pupils	Mean weight
Class A	boys	16	42.6 kg
	girls	15	38.5 kg
Class B	boys	14	41.8 kg
	girls	17	37.9 kg

(a) What is the mean weight of the pupils in Class A?

(b) What is the mean weight of the pupils in Class B?

(c) What is the mean weight of all the pupils in the two classes?

 Challenge and extension question

7 There are three numbers: Number A, Number B and Number C. The mean of Number A and Number B is 36. The mean of Number B and Number C is 30, and the mean of Number A and Number C is 39. What is the mean of these three numbers?

9.4 Application of the mean (1)

 **Learning objective** Solve problems involving finding the mean as an average

 Basic questions

1 Complete each statement.

(a) Jason completed 54 mental sums in 1.5 minutes while Sanjit completed 76 mental sums in 2 minutes. Jason did ⬚ mental sums per minute and Sanjit did ⬚ per minute. _____ did it faster.

(b) Two teams had a competition for making paper flowers. There were 9 pupils in Team A and 8 pupils in Team B. Team A made 108 paper flowers and Team B made 100 during the same period of time. The members in Team ⬚ made paper flowers faster.

2 The table shows the results of a 1-minute skipping rope competition between two teams.

Results of 1-minute rope skipping: Team 1

Name	Ming	Jolie	Mike	Lee	Fay
Number of skips	56	44	60	58	47

Results of 1-minute rope skipping: Team 2

Name	John	Tom	Mary	Joan
Number of skips	50	47	59	62

(a) On average, how many skips did each pupil in Team 1 do in one minute?

(b) On average, how many skips did each pupil in Team 2 do in one minute?

(c) Which team did more skips in one minute per pupil on average, and by how many more skips?

3 Car A travelled 238 km in 3.5 hours and Car B travelled 260 km in 4 hours. Which car travelled at a faster speed?

4 Three groups of pupils went on a trip. Group A had 5 members and drank 6 litres of water during the trip; Group B had four members and drank 5 litres of water; and Group C had three members and drank 4 litres of water. Which group drank the most water per person during the trip?

5 A factory's assembly workshop has three teams. Team 1 has 12 workers and can assemble 162 cars in a day. Team 2 has 15 workers and can assemble 195 cars in a day. Team 3 has 16 workers and can assemble 224 cars in a day. Which team's workers are the fastest in assembling the cars?

6 Evans and his parents went hill climbing one weekend. Evans reached the peak in 48 minutes at a speed of 36 m per minute. When he came down the hill using the same path, it took him 32 minutes.

(a) What was Evans' average speed per minute when he came down the hill?

(b) What was his average speed per minute in the entire journey?

Challenge and extension question

7 A sales assistant mixed 2 kg of chocolate bars, 3 kg of mints and 5 kg of lollipops together. If the chocolate bars are £12.80 per kg, the mints are £8.80 per kg and the lollipops are £9.80 per kg, then what is the price of the assorted sweets per kg?

9.5 Application of the mean (2)

 **Learning objective** Solve problems involving finding the mean as an average

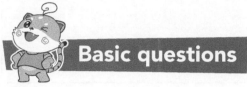 **Basic questions**

1 Complete each statement.

(a) Helen walked 15 m in 25 steps. Her average step was ⬚ m.

(b) Maria can walk 56 steps in a minute on average. If each step is 0.5 m, then Maria can walk ⬚ m in one minute.

(c) Ling walked five times from the school gate to his classroom to estimate the distance. The number of steps he walked each time was 62, 59, 61, 61 and 60, respectively. From the school gate to the classroom, Ling needs to walk about ⬚ steps.

(d) Nancy walked 20 steps and Kim walked 25 steps to cover the same distance of 10 m. The average length of each step Nancy took is ⬚ cm more than that of Kim.

2 Jo wanted to know the distance she could walk in one minute. She tested it six times and the distances were 59 steps, 63 steps, 65 steps, 60 steps, 61 steps and 64 steps, respectively.

(a) How many steps did Jo walk in one minute on average?

(b) If the average length of her step is 45 cm, then how many metres could she walk in 1 minute?

(c) It takes Jo 20 minutes to walk from home to her school. About how long is the distance between her home and the school?

(d) The distance between Jo's home and Oliver's home is about 446.4 m. About how many minutes will it take Jo to walk from her home to Oliver's home?

3 The table shows the amount of after-school reading Martin managed in one week.

Day	Mon	Tue	Wed	Thu	Fri	Sat	Sun
Amount of reading (words)	1800	2000	2100	2400	1600	1500	2600

(a) How many words did Martin read per day on average?

(b) About how many days will it take Martin to finish reading a book of 48 000 words?

(c) Based on the same reading efficiency, about how many words can he read in a year?

4 A car travelled from Place A to Place B for 7 hours at an average speed of 64 km/h. The average speed for the first 4 hours was 70 km/h. What was the average speed for the remaining 3 hours?

5 There are three numbers, A, B and C. The mean of Number A and Number B is 84. The mean of Number B and Number C is 92. Number B is 85. Find the mean of these three numbers.

Challenge and extension question

6 In a Maths test, the mean score of the class was 91.2 marks, but Henry was absent due to illness. After Henry had a make-up test and scored 98 marks, the mean score of the whole class was brought up to 91.4 marks. How many pupils are there in the class in total?

9.6 Application of the mean (3)

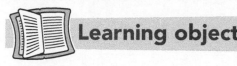

Learning objective Solve problems involving finding the mean as an average

Basic questions

1 Complete each statement.

(a) Number A is 36 and Number B is 42. Number C is 54 and Number D is 72. The mean of the four numbers is ⬚.

(b) The mean of Number A and Number B is 39. The mean of Number C and Number D is 63. The mean of the four numbers is ⬚.

(c) The mean of Number A, Number B and Number C is 44. Number D is 72. The mean of the four numbers is ⬚.

(d) The mean of Number A, Number B, Number C and Number D is 51. The mean of Number A, Number B and Number C is 44. The mean of Number C and Number D is 63. Number C is ⬚.

2 Five boxes of biscuits weigh 240 g, 300 g, 360 g, 380 g and 420 g, respectively.

(a) What is the mean weight of these boxes of biscuits?

(b) After one box of biscuits was eaten, the mean weight of the remaining biscuits per box is 330 g. What was the weight of the box of biscuits that was eaten?

3 In a Maths test, the mean score of Ahmed, Peter and Jane was 92 marks, and the mean score of Cindy and James was 95 marks.

(a) What was the mean score of the five pupils?

(b) If James scored 4 more marks than Cindy, Cindy scored 2 more marks than Peter, and Peter scored 1 more mark than Jane, then how many marks did Ahmed score?

4 In a skipping rope competition, Simon did 135 skips, 146 skips and 150 skips in his first three attempts. To achieve an average of 148 skips, how many skips must Simon do in the fourth attempt?

5 In an end-of year-test, the mean score of a class of 40 pupils was 92.5 marks, with two pupils absent due to illness. In their make-up test, the two pupils scored 95 and 100 marks respectively. What was the new mean score of the class?

6 The distance between Place A and Place B is 3600 km. It takes an aeroplane 4.5 hours to fly from Place A to Place B, and 5.1 hours to return from Place B to Place A. How many kilometres does the aeroplane fly between the two places in one hour on average?

7 After a mid-term test, Marvin found that the mean score of his whole class was 88.4 marks. Later on, he found a mistake. He misread one score of 89.5 as 80.5. After recalculation, the mean score was 88.7 marks. How many pupils were there in the class?

Challenge and extension question

8 Mr Smith was tasked to make some curtains for his company. He made 128 curtains on the first day, 156 curtains on the second day and 164 curtains on the third day. The number of curtains he made on the fourth day was 14 more than the mean number of the curtains he made in these four days. How many curtains in total did Mr Smith make in the four days?

9.7 Pie charts

Learning objective Interpret and construct pie charts to solve problems

Basic questions

1 The figure shows a circle with centre O and radius OA.

(a) Use a protractor to draw ∠AOB = 38°, ∠AOC = 98°, ∠AOD = 170° and ∠AOE = 260° where points B, C, D and E are all on the circumference of the circle.

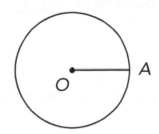

(b) Calculate and then complete the table below. The first one has been done for you.

	∠BOC	∠COD	∠DOE	∠EOA
Measure	60°			
Fraction or percentage of 360°	$\frac{1}{6}$			

2 True or false? (Put a ✓ for true and a ✗ for false in each box.)

(a) Any angle can be classified into one of the following categories: acute angles, right angles, obtuse angles, straight angles and full angles.

(b) Different statistical graphs, charts and tables can be used to represent data and make it easier to understand.

(c) A pie chart represents data using a circle divided into different parts, with each part representing a percentage of the total quantity. ☐

(d) There is always only one way to represent data in statistics. ☐

3 The pie chart below represents the data collected from all the Year 6 pupils in a school about their favourite sports.

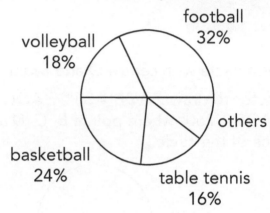

(a) What sport do the greatest number of pupils prefer? _____

(b) What percentage of the pupils preferred volleyball? ☐

(c) What percentage of pupils preferred the other sports? ☐

(d) Given there are 400 pupils in the year group, complete the following table.

	Basketball	Football	Table tennis	Volleyball	Others
Number of pupils favouring					

4 A supermarket carried out a survey of 100 customers on how they travelled to the shop. The result showed that 45% of customers drove, 15% took public transport, the number of customers walking to the supermarket was $\frac{2}{3}$ that of the customers driving, and the rest took the supermarket's special shuttle bus service for customers.

(a) Find the number and percentage of customers who walked to the supermarket.

(b) Find the number and percentage of the customers who got the shuttle bus service to the supermarket.

(c) Construct a pie chart based on the available information.

Challenge and extension question

5 The pie chart below shows the amount of money donated by employees of a company to a disaster relief fundraising event. There were 10 people who donated £10. Answer the following questions.

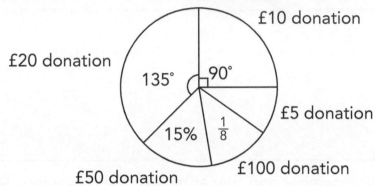

£10 donation

£20 donation

135°

90°

£5 donation

15%

$\frac{1}{8}$

£100 donation

£50 donation

(a) How many people donated?

(b) How many people donated £5?

(c) How much was the average amount each employee donated?

Chapter 9 test

1 Complete each statement.

(a) In three Maths tests, Alvin scored 94 marks, 92 marks and 99 marks.

His mean score was ☐ marks.

(b) The heights of six pupils are 150 cm, 143 cm, 149 cm, 155 cm, 161 cm

and 157 cm. Their average height is ☐ cm. (Keep the answer to one decimal place.)

(c) Number A is 30, which is 6 less than Number B and twice Number C.

The mean of Number A, Number B and Number C is ☐.

(d) The mean of three numbers is 15.6. If 13.2 is included as a fourth

number, then the mean of the four numbers is ☐.

(e) The mean of Number A, Number B and Number C is 48. The mean of

Number A and Number B is 49. Number C is ☐.

(f) A nursery received 15 kg of fruit in the first delivery. In the second delivery, the amount of fruit it received was 1.2 times the amount it received in the first delivery. The fruit was shared equally by six classes

of pupils. Each class received ☐ kg of fruit.

(g) The mean of five numbers is 60. If the mean of the three least numbers is 44 and the mean of the three greatest numbers is 72, then the number

in the middle place, from the least to the greatest, is ☐.

(h) Josh walked five steps and measured their lengths. The results were 46 cm, 48 cm, 48 cm, 50 cm and 52 cm. The mean length of his steps

was ☐ cm.

(i) The mean of Number A, Number B and Number C is 28. The sum of Number D and Number E is 64. The mean of these five numbers is

☐.

(j) The mean score of three pupils in an English test is 85 marks. The

fourth pupil must score ☐ marks so that the mean score of the four pupils is 87 marks.

(k) Avon misread the weight of one of five bags of food as 30 kg. Therefore, he got the mean weight of each bag as 60 kg, while the actual mean weight should be 70 kg. The weight of the bag Avon misread was [____] kg.

(l) An electrical shop sold 105 television sets in the first week and 147 sets in the second week. On average, it sold [____] television sets per week and [____] television sets per day in the two weeks.

(m) It took a delivery man 5.5 hours to travel from Place A to Place B at a speed of 18 km per hour. On his return, he travelled 22 km per hour. The average speed the delivery man travelled for the whole journey was [____] km per hour.

(n) Lynn took three Science tests. She scored 90 marks in the first test and 98 marks in the second. Her score in the third test is 2 marks higher than the mean score of the three tests. Lynn scored [____] marks in the third test.

(o) There are three numbers, O, $\triangle 8$ and $\square 69$ in which O, $\triangle$ and $\square$ stand for three digits from 1 to 9. The mean of the three numbers is 152. The sum of the three digits is [____].

(p) All the books in a school library are classified into three categories, A, B and C, as shown on the pie chart. If there are 9600 books in the library, then there are [____] books, [____] books and [____] books in Category A, B and C respectively.

2 Multiple choice questions. (For each question, choose the correct answer and write the letter in the box.)

(a) A clothing store sold 360 sets of clothes on the first day and 450 sets on the second day. On the third day, it sold 160 sets in the morning and 180 sets in the afternoon. How many sets of clothes did it sell every day on average? The correct expression is ☐.

A. $(360 + 450 + 160 + 180) \div 2$ B. $(360 + 450 + 160 + 180) \div 3$

C. $(360 + 450 + 160 + 180) \div 4$ D. $(360 + 450 + 160 + 180) \div 6$

(b) Tom practised typing in one week. The amount of time he practised each day was 200 minutes, 160 minutes, 190 minutes, 160 minutes, 190 minutes, 160 minutes and 200 minutes, respectively. How much time did he practise per day on average? The correct expression is ☐.

A. $(200 + 160 + 190) \div 3$

B. $(200 \times 3 + 160 \times 2 + 190 \times 2) \div 7$

C. $(200 \times 2 + 160 \times 2 + 190 \times 3) \div 7$

D. $(200 \times 2 + 160 \times 3 + 190 \times 2) \div 7$

(c) A Year 6 class has four groups. In a Maths test, the mean score of Group 1 and Group 2 is 92 marks, the mean score of Group 2 and Group 3 is 91 marks, and the mean score of Group 3 and Group 4 is 91.5 marks. The correct expression for the mean score of the whole class is ☐.

A. $(92 + 91 + 91.5) \div 3$ B. $(92 + 91) \times 2 \div 4$

C. $(91 + 91.5) \times 2 \div 4$ D. $(92 + 91.5) \times 2 \div 4$

(d) The ratio of the number of boys to that of girls in a class is 1 : 2. The mean height of the boys is 154 cm and the mean height of the girls is 148 cm. The mean height of the whole class is ☐.

A. 149 cm   B. 150 cm   C. 152 cm   D. uncertain

(e) Three lorries delivered 68 tonnes of goods in total in three trips and another five lorries delivered 84 tonnes of goods in total in four trips. How many tonnes of goods did each lorry deliver in each trip on average? The correct expression is ☐ .

A. (68 + 84) ÷ 2

B. (68 ÷ 3 ÷ 3 + 92 ÷ 5 ÷ 4) ÷ 2

C. (68 + 84) ÷ (3 × 3 + 5 × 4)

D. (68 + 84) ÷ (3 + 5) ÷ (3 + 4)

(f) The pie chart on the right shows how Jay spent his income in a month for three different purposes. If his income in that month was £3600, then the amount he spent on purpose B was ☐ ?

A. £1150

B. £1200

C. £1300

D. uncertain

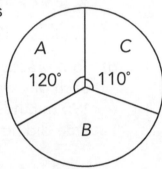

3 A factory has three workshops. There are 147 workers in the first workshop. The second and third workshops each have 153 workers. How many workers are there in each workshop on average?

4 There are two indoor sports rooms. Room A has 120 pupils with an area of 50 m². Room B has 150 pupils with an area of 60 m². Which room is more crowded? Give your reason.

5 The table shows the amount of time Pat spent doing housework in a week.

Day	Mon	Tue	Wed	Thu	Fri	Sat	Sun
Time (minutes)	30	28	15	32	23	40	32
% of the total							

(a) How much time did Pat spend doing housework per day on average in the week?

(b) Find the percentage of the total time Pat spent doing housework on each day in the whole week, and complete the table.

(c) Construct a pie chart based on the available information, indicating the percentage of the time Pat spent each day doing housework.

6 A farmer recorded the amounts of grass six cows ate in a summer month, as shown in the table.

Cow	Cow 1	Cow 2	Cow 3	Cow 4	Cow 5	Cow 6
Amount of grass (kg)	300	350	330	350	350	300

(a) On average, what amount of grass did each cow eat in the month?

(b) About how many kilograms of grass will each cow eat in the whole summer (3 months)?

7 John walked 10 steps four times and recorded the distance he covered each time: 4.2 m, 4.8 m, 4.6 m and 4.8 m, respectively.

(a) What was the average length of a single step?

(b) To find the distance from his home to the gate of his community centre, he walked 5 times between the two places, and it took him 120 steps, 136 steps, 128 steps, 130 steps and 126 steps, respectively. What is the distance between John's home and the gate approximately?

(c) John walked 8500 steps from his home to a local shop. What is the distance (in km) between his home and the shop?

8 A road maintenance team was tasked to repair a road. It repaired 36 m of the road per day on average in the first 12 days and 630 m in total in the next 18 days. What was the length of road that the team repaired per day on average?

9 A toy factory made 2400 toy lorries in January; 200 more than in February. The number of toy lorries made in March was the same as in January. In April, the factory made 600 more than in February. How many toy lorries did the factory make per month on average in the four months?

10 The table shows the mean scores of three Year 6 classes in an English test and the numbers of pupils in these classes.

Class	Class 1	Class 2	Class 3
Number of pupils	32	28	30
Mean score (marks)	90	91.5	88

(a) What is the total score of each class?

(b) What is the mean score of the three classes on average?

(c) What is the mean score of all the pupils in the three classes?

11 Two Year 6 teams had a basketball shooting competition. The table shows the result of each member.

	Member no.	One	Two	Three	Four	Five	Six	Seven	Eight	Nine	Ten
Team A	No. of goals	12	8	10	9	11	9	8	6	15	16
Team B	Member no.	One	Two	Three	Four	Five	Six	Seven	Eight	Nine	
	No. of goals	8	9	12	11	5	10	12	9	14	

(a) How many goals did each member make on average in Team A and Team B, respectively?

(b) How many goals did each member make on average in the two teams altogether? (Give your answer to one decimal place.)

12 Two groups of workers are tasked to produce toys of the same type. On average, everyone is to produce 120 toys. There are 15 people in Group 1 and everyone in Group 1 is to produce 128 toys on average. If everyone in Group 2 is to produce 115 toys on average, how many people are there in Group 2?

Chapter 10 General revision*

10.1 Numbers and their operations (1)

 Learning objective Solve problems involving addition, subtraction, multiplication and division

*This chapter is for the purpose of general revision and it contains some more challenging problems.

 Basic questions

1 Complete each statement.

(a) In 0.8, 3, $\frac{6}{5}$, −7, −1.1, 15, 0, −$\frac{1}{2}$, 0.5 and $\frac{8}{10}$, there are [ ] whole

numbers, [ ] fractions, [ ] decimal numbers, [ ] negative

numbers and [ ] improper fraction(s). The two equal numbers are

[ ] and [ ].

(b) 18.058 consists of [ ] ones and [ ] thousandths. The

difference of the place values of the two 8s in 18.058 is [ ].

(c) The number consisting of 3 hundreds, 4 tens and 5 hundredths is

[ ]. It is a decimal number with _____ decimal

places. If the digit 4 is to appear in the tenths place, the decimal

point should be moved _____ digital places to

the _____.

(d) Fill in the ◯ with >, < or =.

1.357 ◯ 1.537   0.12 ◯ −9.9   3.2̇8 ◯ 3.28̇

(e) The sum of any 10 consecutive whole numbers must be an

_____ number. (Choose: 'odd' or 'even'.)

(f) If a decimal number with three decimal places is rounded to 1.25, to the nearest hundredth, then the greatest possible value of the number

is [ ] and the least possible value is [ ].

(g) After a whole number is inserted with a decimal point, the sum of the decimal number obtained and the whole number is 260.58. The whole

number is [ ].

(h) Using 3, 4, 5 and a decimal point, you can form [ ] different

decimal numbers. The sum of these decimal numbers is [ ].

2 True or false? Put a ✓ or ✗ in each box.

(a) Fractions can be classified into two categories: proper fractions

and improper fractions. [ ]

(b) Dividing a whole into ten parts, one part is one tenth of

the whole. [ ]

(c) A non-zero digit in different value places of a number represents

different values. [ ]

(d) In a decimal number, the further a digit is from the decimal point on

the right, the greater the place value it stands for. [ ]

3 Complete each statement.

(a) Because 27 ÷ 9 = 3, we say that [ ] is divisible by [ ].

(b) In 1, 2, 3, 6, 9, 12, 15 and 24, there are [ ] factors of 6 and [ ]
multiples of 6.

(c) The difference between the least multiple and the least factor of 248

is [ ].

(d) The least two-digit number that is divisible by 2, 3 and 5 is [ ].

(e) In the positive integers that are less than 10, the number that is both an odd number and a composite number is ☐.

(f) The sum of all the prime numbers within 10 is ☐.

(g) The prime factors of 24 are ☐.

4 Multiple choice questions. (For each question, choose the correct answer and write the letter in the box.)

(a) In 30.040, 30.400, 30.04, 34.00, 3.04 and 30.0400, there are ☐ numbers that are equal.

    A. 2   **B.** 3   **C.** 4   **D.** 5

(b) Put 0.3, −0.6, 0.03, −0.3 and −0.36 in order, from the greatest to the least. The 4th number is ☐.

    A. −0.6   **B.** 0.03   **C.** −0.3   **D.** −0.36

(c) If $a \div 2 = b$ ($a > 0$, $b > 0$), then $a \div b =$ ☐.

    A. 2   **B.** 1   **C.** 0.5   **D.** uncertain

(d) When Ben was adding two decimals, he mistakenly swapped the tenths place in one of the decimals to its hundredths place. The difference between the sum he obtained and the correct sum could be ☐.

    A. 0.24   **B.** 0.63   **C.** 0.75   **D.** 0.9

Challenge and extension questions

5 How many 3-digit numbers are there satisfying both of the following conditions?

(a) They are multiples of 5.

(b) The sum of the three digits is 10.

6 There are 40 consecutive even numbers and the greatest number is four times the least number. What is the sum of these 40 even numbers?

10.2 Numbers and their operations (2)

Learning objective Calculate using the four operations and brackets

Basic questions

1 Look at the order of operations carefully and then write the answers.

(a) $0.3 + 0.3 \div 6 - 0.3 \div 3 =$

(b) $(0.3 + 0.3) \div 6 - 0.3 \div 3 =$

(c) $10 + 5 \div (8 - 4) \div 5 =$

(d) $(10 + 5 \div 8 - 4) \div 5 =$

2 Complete each statement.

(a) The sum of two numbers is 10. If one addend increases by 1.5 and the other addend decreases by 1.5, then the new sum is   .

(b) The difference of two numbers is 10. If the minuend increases by 1.5 and the subtrahend decreases by 1.5, then the new difference is   .

(c) The product of two numbers is 10. If one factor is multiplied by 2 and the other factor is divided by 4, then the new quotient is   .

(d) The quotient of two numbers is 10. If the dividend is divided by 2 and the divisor is multiplied by 4, then the new quotient is   .

3 Work these out step by step.

(a) $24 \div (48 \times 0.5) - 0.98$

(b) $-8 \times (1.4 - 1.15) + 7.5 \times 18$

(c) $(3.9 - 3.6 \div 4) \times 1.2$

(d) $[(7.2 - 5) \div 2 + 9.9] \times 0.6$

(e) $(3.24 + 5.8 + 1.76) \div (0.4 \times 15)$

(f) $(13.5 + 4.5) \times [6.6 - (2.9 + 3.7) \div 12]$

4 Calculate smartly with the following. Show your working.

(a) 28 + 2.8 × 37 + 2.8 × 53

(b) 5.37 − 4.72 + 4.63 − 3.28

(c) 164 × 8 ÷ 82 × 12.5

(d) 5.4 ÷ (3 × 2.4 + 3 × 2.6)

(e) 15 × 3.4 + 150 × 0.46

(f) 3.9 × 78 − 3.9 × 22 − 56 × 2.9

5 Calculate: $2.5 \times 24 \div 14 + 48 \div 21 \times 2.5$

6 Calculate: $3333 \times 66.66 + 44.44 \times 9999$

10.3 Numbers and their operations (3)

Learning objective Solve problems involving the calculation and conversion of units of measure

Basic questions

1 Convert these units of measure.

(a) 1.5 km = [   ] m

(b) 34 cm = [   ] mm

(c) 0.9 m² = [   ] cm²

(d) 280 000 m² = [   ] km²

(e) 12 cm³ = [   ] mm³

(f) 550 ml = [   ] l

(g) 8.2 kg = [   ] g

(h) 4700 g = [   ] kg

(i) 2.5 days = [   ] hours

(j) 480 seconds = [   ] minutes

(k) 6 km 6 m = [   ] km

(l) 35 020 g = [   ] kg [   ] g

(m) 7.05 m³ = [   ] m³ [   ] cm³

(n) 9.6 m² + 880 cm² = [   ] m²

(o) 50 500 ml = [   ] m³

(p) 3 hours 48 minutes = [   ] minutes

2 Complete each statement.

(a) The length of one step is approximately 45 cm. The total length of 16 steps is about [   ] cm.

(b) The base of a triangle is 125 cm, which is 45 cm more than its height. The area of the triangle is [   ] m².

(c) If you pour 100 litres of juice into 350 bottles, on average each bottle will contain [ ] millilitres at most. (Round your answer to the nearest whole number.)

(d) A cake weighs 280 g. A big cardboard box contains 150 such cakes. The total weight of the cakes in the box is [ ] kg.

(e) One day, Tom went to a science museum. He left home at 8:00 and came back at 11:00 in the morning. The time he spent at the science museum was 1.5 times the time he spent travelling there and back.

How many minutes did he spend at the science museum? [ ]

3 Solve these problems.

(a) If 5 kg of sugar cane can produce 480 g of sugar, then how many kilograms of sugar can be produced with 1500 kg of sugar cane? How many tonnes of sugar cane is needed to produce 1.2 tonnes of sugar?

(b) The total length of the edges of a cuboid paper box is 280 cm. The length, width and height of this cuboid box are all different multiples of 10 in centimetres. How many cubic metres is the volume of the cuboid paper box?

(c) A square piece of paper with a length of 1 m is cut into small squares with a length of 5 cm. If these small squares are put in a row to form a rectangle, how many metres is the length of the rectangle?

(d) Three pupils can type 5400 words in 15 minutes. At this speed, if two more pupils join in the typing, how many words can they type in 1.4 hours?

Challenge and extension question

4 Peter is running from west to east at 200 m per minute on a path parallel to a railway, where a train is travelling in the opposite direction at 48 km per hour. If it takes 24 seconds for the train to pass Peter completely, what is the length of the train? If another train, 200 m long, is travelling in the same direction as Peter at 60 km per hour on a parallel railway, how long will it take for the two trains to pass each other completely?

10.4 Numbers and their operations (4)

Learning objective Solve problems involving addition, subtraction, multiplication and division

Basic questions

1 Complete each statement.

(a) Given $8.24 \times 525 = 4326$, we know that $0.824 \times 525 = \boxed{}$,

$43.26 \div 824 = \boxed{}$.

(b) If $[8.7 + (\boxed{} - 7.15) \times 6] \div 12 = 1.15$, then $\boxed{} = \boxed{}$.

(c) If the sum of three different positive integers is 20, then the least

possible product of the three numbers is $\boxed{}$ and the greatest

possible product of these numbers is $\boxed{}$.

(d) If the product of the sum of two whole numbers multiplied by their difference is 77, then the product of these two whole numbers is

$\boxed{}$ or $\boxed{}$.

(e) Think about the division of a 3-digit number by 59. To have the greatest possible sum of the quotient and the remainder, the 3-digit

number is $\boxed{}$.

2 Solve these problems.

(a) The pupils in Year 3, Year 4 and Year 5 in a school donated books to a charity. Year 3 donated 156 books, 18 more than Year 4. The number of books Year 5 donated was 1.5 times the number of the books donated by Year 3 and Year 4. How many books did Year 5 donate?

(b) Mr Lee was tasked to make a batch of spare parts. In the morning, he completed half of the task after working 3.5 hours at 18 spare parts per hour. In the afternoon, he continued to work but at 14 spare parts per hour and completed his task. How many hours did it take Mr Lee to complete the task?

(c) A box of nails weighs 750 g. After taking out 54 nails, it weighs 615 g. Given that all the nails in the box are the same, how many nails were there in the box at first?

(d) A factory received a delivery of coal. It planned to use 5 tonnes per day so the coal could last 36 days. After an improvement in the plan, it had actually saved 0.5 tonnes of coal per day. How many more days can the coal last than the original plan?

(e) Ming's average score of Mathematics, Chinese and English is 94. His Maths score is 3 marks more than the average while his Chinese score is 5 marks fewer than the average. What is his English score?

(f) Two cars started travelling towards each other at the same time from two places which are 900 km apart. Car A travelled at 60 km per hour. When Car A reached the midpoint of the whole journey, it was 30 km away from Car B. Find the speed of Car B.

(g) Joan walked from home to a community centre at 65 m per minute. After 18 minutes, her mother found that Joan had left a document at home. She cycled to catch up with Joan at 195 m per minute. Given that the distance between the community centre and Joan's home is 1800 m, how far was she away from the community centre when Joan's mother caught up with her?

Challenge and extension question

3   Two cars started travelling towards each other from two places at the same time. After 2.4 hours, the two cars had not met and were 14 km apart from each other. After 2.8 hours, they were 42 km away from each other. Find the distance between the two places.

10.5 Numbers and their operations (5)

Learning objective Solve problems involving fractions, decimals and percentages

Basic questions

1 Multiple choice questions. (For each question, choose the correct answer and write the letter in the box.)

(a) A road is being built. On the first day, $\frac{2}{5}$ of the whole road was built. On the second day, $\frac{1}{3}$ of the whole road was built. After the two days, ☐ of the whole road was left unfinished.

    A. $\frac{3}{15}$ **B.** $\frac{4}{15}$ **C.** $\frac{2}{15}$ **D.** $\frac{11}{15}$

(b) If the price of an item is £5832 after a discount of $\frac{1}{10}$, then the original price was ☐.

    A. £6413   **B.** £6480   **C.** £5248.8   **D.** £7128

(c) The greatest integer less than $\frac{25}{4}$ is ☐.

    A. 5   **B.** 6   **C.** 7   **D.** 8

(d) Given that $a = 1 + \frac{8}{9}$, $b = 2 - \frac{2}{9}$, and $c = 3 - 1\frac{5}{8}$, the right order of a, b and c in the following is ☐.

    A. $a < b < c$ **B.** $a < c < b$ **C.** $b < a < c$ **D.** $c < b < a$

(e) If the number of boys is $\frac{5}{4}$ of the number of girls in the class, then the ratio of the number of girls to the number of the whole class is ☐.

    A. $\frac{1}{5}$ **B.** $\frac{5}{4}$ **C.** $\frac{4}{9}$ **D.** $\frac{5}{9}$

(f) If you convert $3 \times 2.5 = 5 \times 1.5$ to a proportion, the correct one is ☐.

    A. $\frac{3}{2.5} = \frac{5}{1.5}$ **B.** $3:5 = 1.5:2.5$

    C. $3:1.5 = 2.5:5$ **D.** $3:5 = 2.5:1.5$

2 Complete each statement.

(a) 20% of ☐ is 30.

(b) If $\frac{64}{256}$ is converted to a decimal, the result is ☐ ; if it is

converted to a percentage, the result is ☐ . 6.35 converted to

a fraction in its simplest form is ☐ .

(c) There are ☐ fractions in the simplest form with the denominator 14.

(d) A steel tube of $\frac{5}{6}$ m long is cut into 5 equal pieces. The length of each piece is ☐ m. The length of each piece is ☐ of the whole piece.

(e) Fill in the blank with a mixed fraction: 200 minutes = ☐ hours.

(f) Put the numbers 0.83, 0.8$\dot{3}$ and $\frac{41}{49}$ in order, starting from the least:

_____ .

(g) Converting the fraction $\frac{5}{12}$ to a decimal, correct to three decimal places, the result is ☐ .

(h) If the numerator of a fraction is 51 and after simplification it is $\frac{3}{4}$, then the fraction before simplification was ☐ .

(i) Given that the number of boys in a class is $\frac{2}{5}$ of the number of girls, then the number of girls is ☐ of the whole class.

(j) Given $x:y = \frac{1}{4}:\frac{1}{2}$, $y:z = 1.2:0.6$, then $x:y:z =$ _____ .

3 Calculate the answers.

(a) $\frac{3}{14} \times \frac{5}{9}$

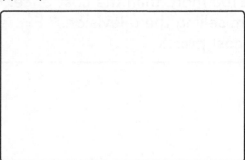

(b) $\frac{1}{5} \times \frac{10}{11}$

(c) $\frac{3}{14} \div 2$

(d) $\frac{1}{3} \times \frac{9}{16} + \frac{7}{8} \div 5$

(e) $4\frac{1}{2} + 4 \times \frac{7}{8} + \frac{3}{8} \div 12$

(f) $6 \times \left(\frac{1}{2} - \frac{1}{3}\right) + \left(\frac{5}{8} + \frac{3}{16}\right) \times \frac{8}{23}$

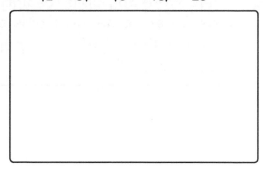

4 A canteen bought a barrel of oil weighing 100 kg. It used $\frac{2}{5}$ of the oil on the first day. On the second day it used $\frac{2}{3}$ of the remaining oil. How many kilograms of oil were left after the second day?

5 A shop bought ten televisions at a cost price of £200 per TV. Two TVs were sold at a price that was 20% more than the cost price. The remaining TVs were sold at a price that was £100 more than the cost price. How much profit did the shop make from selling the televisions? Express the profit as a percentage of the total cost price.

Challenge and extension question

6 There were 36 pupils reading in the school reading room. $\frac{4}{9}$ of them were girls. Later, some more girls entered the reading room. The number of girls was then $\frac{9}{19}$ of all the pupils in the room. How many girls entered the reading room?

10.6 Equations and algebra (1)

Learning objective Use equations to solve problems

Basic questions

1 Complete each statement.

(a) There are two consecutive natural numbers. If the first number is a,

then the second can be written as _____.

(b) If the base of a triangle is 6 cm and the height is y cm, then the area of

the triangle is _____ cm².

(c) John walks 4 km per hour and Andrew walks 5 km per hour. If they have walked x hours at the same time, then they have walked

_____ km in total.

(d) There were a litres of water at the beginning. After 8 days of use, there were b litres of water left. The average usage of the water was

_____ litres per day during the first 8 days.

(e) There are x chickens and y rabbits in the same hutch. The number

of chicken's feet is _____ fewer than the number of rabbits' feet.

(f) A car was travelling from City A to City B at v km per hour. After t hours, the car reached the midpoint of its journey. The distance

between the two cities is _____ km.

(g) Number a is multiplied by number b. If number a increases by 3 and number b is unchanged, then the new product will be

_____ greater than the original product.

(h) Given that $a = 12$, then $2a - a^2 \div 10 = \boxed{}$.

(i)   Half of number A is 1.5 times number B. If number B is 4.2, then

number A is ☐ .

(j)   If 60 is split into the sum of three numbers, A, B and C, and

A + 3 = B × 3 = C, then A = ☐ , B = ☐ , C = ☐ .

2 True or false? Put a ✓ or ✗ in each box.

(a)   If two identical squares with side length of a are used to form a

rectangle, the perimeter of the rectangle is 8a. ☐

(b)   A water pump shifted x litres of water in 2 hours. If it works 5 more
hours with the same efficiency, then it can shift 3.5x litres of water in

total. ☐

(c)   A car travelled from City A to City B in 7 hours at a km per hour.
Given it travelled b km in the first 4 hours, the speed of the car in the

remaining 3 hours was (7a − 4b) ÷ 3 km per hour. ☐

3 Multiple choice questions. (For each question, choose the correct answer
and write the letter in the box.)

(a)   In the following equations, the correct one is ☐ .

   A. $a^3 = a \times 3$ **B.** $3(x + y) = 3x + y$

   C. $b + b = 2b$ **D.** $2x + y = 2y + x$

(b)   Bill is x years old and his father is y years old. His father is k years older
than him. After m years, how many years older is Bill's father than him?

The correct equation is ☐ .

   A. $x - y = k$ **B.** $x - y = k + m$

   C. $y - x = k$ **D.** $y - x = k + m$

(c)   Each side of a square flowerbed has n pots of flowers (including one

pot of flowers at each corner). There are a total of ☐ pots of flowers
around the flowerbed.

   A. $4n - 4$ **B.** $4n + 4$ **C.** $4n - 8$ **D.** $4n$

(d) There are three numbers A, B and C. If the mean of A and B is x, the mean of B and C is y, and the mean of A and C is z, then the mean of A, B and C can be expressed as ☐.

   A. $(x + y + z) \div 3$ **B.** $(2x + 2y + 2z) \div 3$

   C. $(2x + 2y + 2z) \div 2$ **D.** $(x + y + z) \div 2 \div 3$

4 Two groups of people are sending their representatives to a committee. The number of people in the first group to that in the second group is in the ratio of $4 : 5$.

(a) If there are 60 people in the first group, how many people are there in the second group?

(b) If the first group sends 8 representatives and the second group sends 12 representatives to the committee, are the numbers of representatives sent to the committee in proportion to the numbers of people in the two groups? If so, give your reason. Otherwise, find which group is underrepresented and which group is overrepresented in terms of the ratio of the numbers of people in the two groups.

Challenge and extension question

5 If the solution for x to equation $ax + 20 = 104$ is equal to the solution to $3x - b = 8$, and $ab = 60$, then what are the values of a, b and x?

10.7 Equations and algebra (2)

Learning objective Solve equations

Basic questions

1 Solve these equations. (Check the answers to the questions marked with *.)

(a) $3.9 - 2x = \frac{3}{2}$

(b) $4x + 0.9 = 1.7$

(c) $2(2.8 + x) = 16.4$

(d) $(10.2 - x) : 3 = 2.9$

*(e) $2.5x = 10 + \frac{1}{2}x$

*(f) $3x : 15 = 4 : 5$

(g) $4x + 3.9 \div 1.3 = 6.8$ *(h) $15x - (0.9 - 0.3) = 12$

2 Write the equation and find the solution for each question.

(a) The product of 1.5 and 6 is 0.4 greater than a number. Find this number.

(b) The result of subtracting the quotient of 2.2 divided by 4 from $\frac{1}{2}$ of a number is 5. Find the number.

(c) The mean of Number A and Number B is 36. Number A is 1.4 times Number B. Find Number A and Number B.

(d) Subtracting 4.8 from 6.5 times a number is equal to 5.5 times the number. Find the number.

(e) The sum of 2 times a number and 4 times the number is 9.6 less than 7 times the number. Find the number.

3 The sum of three consecutive whole numbers is 120. Find the three numbers.

Challenge and extension question

4 3 times the sum of a number and 5.4 is equal to 5 times the difference between the number and 1.8. Find the number.

10.8 Equations and algebra (3)

 Learning objective Use equations to solve problems

 Basic questions

Write the equation and find the solution to each problem.

1 At a school, there are 237 pupils in Year 6, which is 24 pupils fewer than 1.5 times the number of pupils in Year 5. How many pupils are there in Year 5?

2 A clothing factory plans to make 1960 sets of clothes. It has made 620 sets and the rest of the clothes should be made in 4 days. How many sets of clothes are to be made each day on average?

3 In the school table tennis team, the number of boys is 1.2 times the number of girls. There are 6 more boys than girls. How many boys and girls are there in the table tennis team?

4 There are 3 kinds of fruit trees in an orchard. There are 360 apple trees, which is 40 trees more than the pear trees. The pear trees are 20 fewer than twice the number of the peach trees. How many peach trees are there in the orchard?

5 Some oranges are shared among children. If each child is given 3 oranges, there are 18 oranges left. If each child is given 5 oranges, it is just enough and no more oranges are left. How many children are there? How many oranges are there in total?

6 Tom went to a bookshop with £100. He bought four storybooks at the price of £13 per book. Later, he bought several graphic novels at £15 per book. He received £3 in change. How many graphic novels did he buy?

7 John has a bicycle. The perimeter of its front wheel is 250 cm and that of the rear wheel is 180 cm. John cycled from Place A to Place B. The rear wheel went round 1001 more times than the front wheel. Find the distance between Place A and Place B.

8 The sum of Number A, Number B and Number C is 112. Number A is 3 times Number B. Number B is 8 more than Number C. Find Number A, Number B and Number C.

9 Car A and Car B started travelling towards each other at the same time from two places 390 km apart. After 5 hours, the two cars met each other. If Car A travelled 2 km faster per hour than Car B, then how many kilometres per hour did Car A and Car B travel?

Challenge and extension question

10 Mary walks to school every day. If she walks 60 m per minute, she will be there 6 minutes late. If she walks 80 m per minute, she will be there 3 minutes early. If Mary wants to be there on time, how many metres should she walk per minute?

10.9 Geometry and measurement (1)

Learning objective Solve problems involving angle properties

Basic questions

1 Complete each statement.

(a) Angles can be classified into acute angles, _____ angles,

_____ angles, _____ angles, _____
angles and full angles.

(b) If two angles of a triangle are 35° and 55°, then it is a

_____ triangle.

(c) In the diagram below, if ∠AOB = 40°, ∠BOC = ∠COD,

then ∠COD = ☐ . If ∠BOE = ∠EOC, then ∠AOE is an

_____ angle.

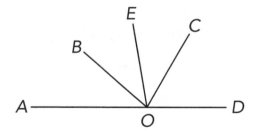

(d) The diagram on the right shows a square.

There are ☐ right-angled triangles

in the diagram.

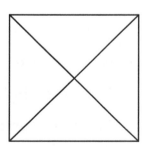

(e) An angle is 75°. If this angle is viewed through a

2 times magnifying glass, it is _____ .

2 True or false? Put a ✓ or ✗ in each box.

(a) The size of an angle is not related to the side lengths of the angle. ☐

(b) When an angle is smaller than 180°, it is either an acute angle or an obtuse angle. ☐

(c) If you divide a line of 100 cm into 10 equal parts, then each part is 10 cm long. ☐

(d) All triangles can be classified by their sides into two categories: isosceles triangles and equilateral triangles. ☐

(e) If one of the angles in a triangle is greater than the sum of the other two angles, then it must be an obtuse-angled triangle. ☐

3 Multiple choice questions. (For each question, choose the correct answer and write the letter in the box.)

(a) In an isosceles triangle, if one of the angles is 50°, then it is ☐ triangle.

    A. an equilateral   **B.** an acute-angled

    C. a right-angled   **D.** an obtuse-angled

(b) In the following shapes, ☐ does not have line symmetry.

    A. a circle   **B.** a rectangle

    C. a parallelogram   **D.** an isosceles triangle

(c) Fold a square piece of paper in half three times, the final figure is ☐ .

    A. a square

    B. a rectangle

    C. an isosceles right-angled triangle

    D. a rectangle or an isosceles right-angled triangle

(d) If the ratio of the three angles in a triangle is 6 : 2 : 1, then the triangle is ☐ .

    A. a right-angled triangle   **B.** an acute-angled triangle

    C. an obtuse-angled triangle   **D.** uncertain

(e) When the minute hand and hour hand on a clock face form a 90° angle, it is ☐ .

    A. 03:30   **B.** 06:15   **C.** 09:00   **D.** 09:30

(f) On a coordinate plane, if point $A(-2, 5)$ is reflected in the x-axis, then the coordinates of its image A' are ☐ .

    A. $(2, 5)$ **B.** $(-2, -5)$ **C.** $(2, -5)$ **D.** uncertain

(g) If point $P(-a, -b)$ is reflected in the y-axis, then the coordinates of its image P' are ☐ .

    A. (a, b) **B.** $(a, -b)$ **C.** $(-a, b)$ **D.** uncertain

(h) If point $A(5, -4)$ is translated 3 units left, then the coordinates of its image A' are ☐ .

    A. $(8, -4)$ **B.** $(2, -4)$ **C.** $(5, -7)$ **D.** uncertain

(i) If point $P(a, b)$ is translated 2 units right and 1 unit up, then the coordinates of the new position the point reaches are ☐ .

    A. $(a + 2, b + 1)$ **B.** $(a - 2, b - 1)$

    C. $(a + 2, b - 1)$ **D.** uncertain

4 Look at the coordinate grid below.

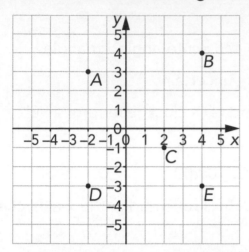

(a) Use coordinates to describe the positions of points *A*, *B*, *C*, *D* and *E*. Which quadrant does each point belong to? Fill in the table below.

	A	*B*	*C*	*D*	*E*
Coordinates					
Quadrant					

(b) Look at the five points. Which three points will form a right-angled triangle when they are connected accordingly? How many pairs of perpendicular lines can be obtained by connecting any two of the five points?

Challenge and extension question

5 Use a circle and five lines to divide a square into different parts. How many parts at most can the square be divided into?

10.10 Geometry and measurement (2)

Learning objective Calculate the areas of parallelograms and triangles

Basic questions

1 Complete each statement.

(a) The base of a triangle is 15 cm and the height is 9 cm. The area of the triangle is ⬜ cm².

(b) The areas of a square and a rectangle are equal. The side length of the square is 60 cm and the width of the rectangle is 40 cm. The perimeter of the rectangle is ⬜ cm.

(c) The base length of a parallelogram is 4 m, which is 5 times the height. The area of the parallelogram is ⬜ m².

(d) The diagonal of a square is 16 cm. The area of the square is ⬜ cm².

2 Multiple choice questions. (For each question, choose the correct answer and write the letter in the box.)

(a) The diagram shows two parallelograms which are partly shaded.

Given *AB* is parallel to *CD*, Area M is ⬜ Area N.

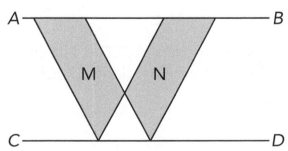

A. greater than
B. equal to
C. less than
D. not comparable with

(b) The correct statement of the following is ☐.

    A. If the areas of two rectangles are equal, then their perimeters must be equal too.

    B. If the areas of two parallelograms are equal, then they must be the same.

    C. The area of a parallelogram is twice the area of a triangle with the equal base and equal height.

    D. Two triangles with equal areas can form a parallelogram.

3 In each figure below, draw the height on the base *BC*.

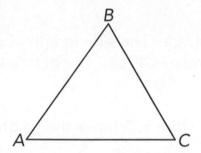 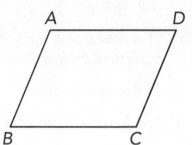

4 The perimeter of a rectangle is 48 cm and the length is 3 cm more than twice the width. Find the area of the rectangle.

5 Look at the diagram. The area of parallelogram *ABCD* is 108 cm². *M* is a point on *AB*, and *CE = EF = FD*. Find the area of triangle *MEF*.

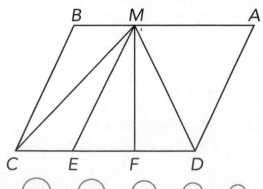

6 In a park, there is a triangular garden with an area of 180 m². The base of the garden is 24 m. Soon the garden is to be expanded with the base increased by 6 m and the height by 3 m. What will be the area of the expanded garden?

Challenge and extension question

7 The diagram shows a composite shape made up of three squares. From left to right, the side lengths are 8 cm, 10 cm and 6 cm respectively. Line AB divides the composite shape into two parts. What is the difference between the areas of these two parts? (Note: A composite shape is made up of two or more shapes.)

10.11 Geometry and measurement (3)

 Learning objective Solve problems involving the areas of shapes

 Basic questions

1 Calculate the areas of the composite shapes shown below (unit: cm).

(a)

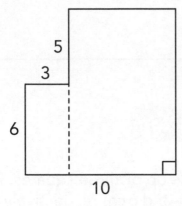

5

3

6

10

(b)

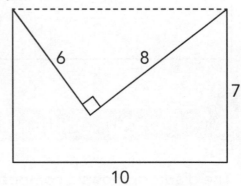

6

8

7

10

Area = _____

Area = _____

2 As shown in the diagram below, the two heights of parallelogram *ABCD* are 15 cm and 18 cm, respectively. *BC* is 24 cm. What is the length of *CD*?

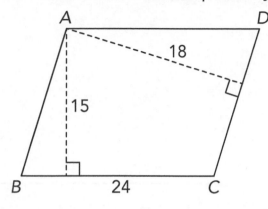

3 The diagram on the right shows a composite shape made up of 5 identical squares with side length 5 cm. What is the area of the shaded part?

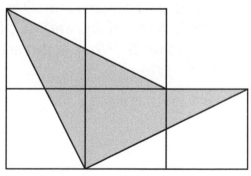

4 Look at the diagram below. The side length of square *ABCD* is 5 cm. *BE* is 12 cm. How much larger is the area of triangle *CEF* than the area of triangle *AFD*?

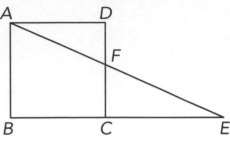

5 Look at the diagram on the right. The side length of square *ABCD* is 9 cm. *E* and *F* are the midpoints of *BC* and *CD*. *BF* and *DE* intersect at point *G*. What is the area of quadrilateral *ABGD*?

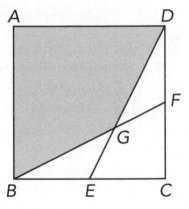

6 On the coordinate plane below, reflect two points $A(-2, -5)$ and $B(2, -5)$ in the x-axis to get points M and N respectively, and then translate both M and N 5 units right to get points P and Q respectively.

(a) Draw and label the six points as indicated and connect the points to get quadrilateral *AMNB* and quadrilateral *APQB*.

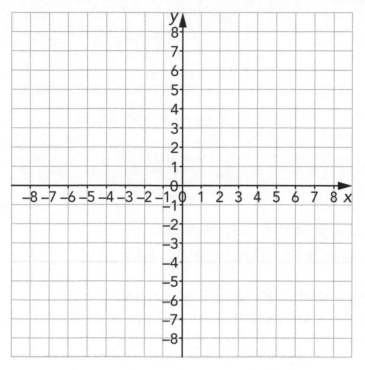

(b) What shape is quadrilateral *AMNB*? What shape is quadrilateral *APQB*?

AMNB = _____

APQB = _____

(c) Find the areas of quadrilateral *AMNB* and quadrilateral *APQB*. Are they the same? Why or why not?

Area of *AMNB* = ☐

Area of *APQB* = ☐

 Challenge and extension questions

7 Point *O* is inside triangle *ABC*. The length of perpendicular lines from point *O* to each side of the triangle is 2 cm and the perimeter of the triangle is 20 cm. What is the area of triangle *ABC*? (Hint: you may draw a diagram to help you.)

8 The diagram below shows a right-angled isosceles triangle *ACB*, *DE // BC*, *BC* = 10 cm and *DE* = 7 cm. What is the area of the shaded region, *BDEC*?

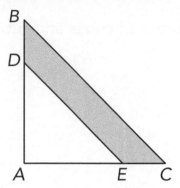

Area = [   ]

10.12 Geometry and measurement (4)

 Learning objective Calculate the volumes of cubes and cuboids

 Basic questions

1 Complete each statement.

(a) The length of a cuboid is 5 cm, the width is 4 cm and the height is 2 cm.

The volume of the cuboid is ☐ cm³.

(b) If the edge length of a cube is 80 cm, then the volume is

☐ cm³.

(c) The edge length of a cube is 1 m. After the surface of the cube is painted red, it is cut into small identical cubes with edge length of

10 cm. There are ☐ such small cubes in total. ☐ of them

have red paint on three sides, ☐ of them have red paint on two

sides and ☐ of them have red paint on one side.

(d) A sealed cuboid container is 8 cm long, 6 cm wide and 5 cm high and has water inside. The water level is 4.8 cm high. If the container is turned over so that its back face becomes the bottom face, then the

water level will be ☐ cm high. If the left face of the container

becomes the bottom face, then the surface of the water will be

☐ cm away from the top of the container.

(e) A cuboid has ☐ faces. If the bottom face of a cuboid is a square,

then a net of the cuboid consists of ☐ identical rectangles and

☐ identical squares.

(f) A net of a square-based pyramid consists of ☐ square (s) and

☐ triangles.

2 The diagram below shows the English letter E. All the measurements are shown in the figure. Find the volume of the letter E. (Unit: cm; drawing not to scale.)

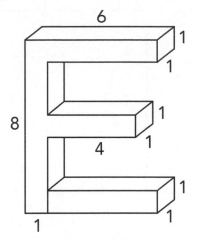

Volume =

3 A room is 5 m long, 4 m wide and 2.8 m high. The owner plans to paint all the walls and the ceiling of the room except the door and windows, which take up a total of 16 m². The cost of paint is £4.50 per m². How much will the paint cost in total?

4 There is a cuboid-shaped barn. The internal length is 16 m and the width is 15 m. If the height of the barn that wheat can be stored up to is 2.5 m and the weight of wheat per cubic metre is 750 kg, then how many tonnes of wheat can be stored in this barn? (Note: 1 tonne = 1000 kg.)

5 A cuboid has two squares on the left and right sides. The perimeter of each square is 20 cm. The sum of the edge length of the cuboid is 72 cm. What is the volume of the cuboid?

Challenge and extension question

6 The length of a cuboid is 1.5 times the height, and the width is twice the height. If the height is increased by 2 m and the length and the width remain unchanged, then the volume is increased by 54 m³. What are the volumes of the original and the enlarged cuboids?

10.13 Statistics (1)

Learning objective Solve statistics problems that include calculating the mean as an average

Basic questions

1 The marks a Year 6 class scored in a Maths test are as follows:

> 90, 88, 95, 78, 100, 93, 95, 89, 98, 91, 91, 100, 83, 96, 75, 86, 99, 100, 65, 73, 96, 98, 100, 93, 97

(a) Use the data above to make a statistics table below.

Marks	90–100	80–89	70–79	60–69	Total
Number of pupils					

(b) Present the data above using a bar chart.

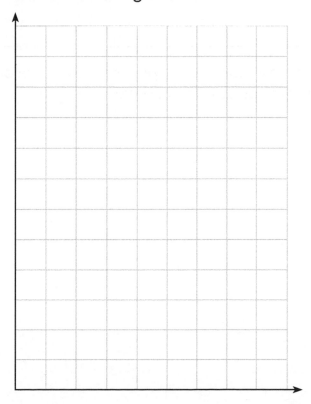

Based on the original data or data presented in the table or bar chart above, answer the following questions.

(c) In which range of marks did the pupils in the class score most?

_____

(d) How many pupils scored 90 or above?

(e) What is the mean score of the class?

2 Three groups of volunteers are planting trees in the spring time. Each person plants 10 trees on average. There are 12 people in the first group, 18 people in the second group and 20 people in the third group.

(a) Express the number of people in each group as a percentage of all the people who volunteered.

(b) In the circle below, construct a pie chart to represent the percentages for the three groups obtained above.

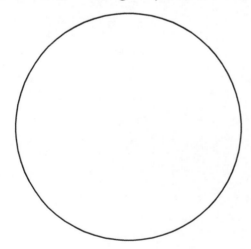

(c) If the first group planted 7 trees per person on average, and the second group planted 9 trees per person on average, then on average how many trees does each person in the third group need to plant in order to plant all the trees?

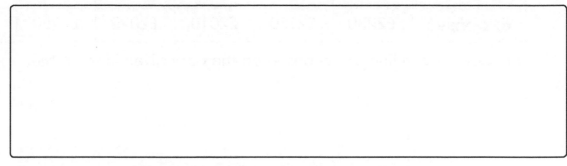

(d) Based on the data given in (c), express the number of the trees planted by each group as a percentage of all the trees planted and then construct a pie chart to represent these percentages.

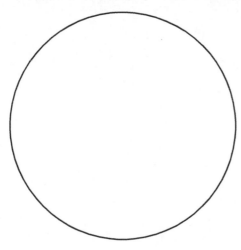

(e) Compare the two pie charts in (b) and (d). Are they the same? Use proportional relationship to explain your conclusion.

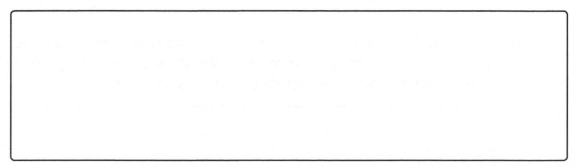

3 The table below shows the monthly expenses of a family in the first six months of 2014, correct to the nearest ten pounds.

Month	Jan	Feb	Mar	Apr	May	Jun
Expenses	£2230	£2150	£2010	£2020	£1950	£1980

(a) Construct a line graph based on the data given in the table.

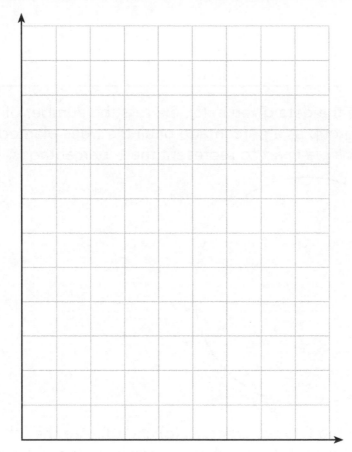

(b) How much were the total expenses of the family in the six months? What was the average per month? What was the average per day? (Give your answers correct to the nearest pound.)

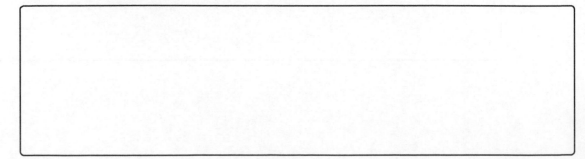

Challenge and extension questions

4 In Question 3 above, what pattern of the family's expenses over the six months did you notice? How would you interpret the pattern you observed? (Explain one or two reasons.)

5 Write down the names of all the statistical tools that you have learned for presenting data, and explain one or two advantages and disadvantages of each tool. (Hint: You may wish to use a table to summarise.)

10.14 Statistics (2)

Learning objective Solve statistics problems that include calculating the mean as an average

Basic questions

1 It took Leo 4 hours to drive from Place A to Place B. He drove 60 km in the first hour, 57 km in the second hour, 55 km in the third hour and 50 km in the last hour. What was his average speed for the whole journey?

2 Joan participated in a singing competition. There were 7 judges. When calculating the scores for a candidate, the highest and lowest scores were not counted, and the mean of the remaining scores was the final score for the candidate. Joan's scores given by the 7 judges were 9.2, 8.7, 9.9, 9.1, 8.5, 9.5 and 9.2. What was Joan's final score?

3 There are two teams in a workshop. The first team has 7 workers and they have made 13 500 spare parts. The second team has 8 workers and they have made 16 200 spare parts. How many spare parts did each team make on average? How many spare parts did each worker make on average?

4 In mid-year exams, Ben scored 98 marks in Maths, which was 7 marks more than his Science and 2 marks more than his English. What was the mean score of his three subjects?

5 A team of 10 pupils participated in a Maths competition. Two of them scored 100 marks, two scored 98 marks, one scored 95 marks, and all the other 5 pupils scored 454 marks in total. What was the mean score of the team in the Maths competition?

6 A school organised an outing. Seven pupils were grouped into a team. They bought 12 small pizzas to share. As Pupil A and Pupil B did not have any money with them, the other 5 pupils shared the cost. The next day, Pupil A and Pupil B paid back £9.60 in total to the other 5 pupils. How much did each pizza cost?

7 The length of a path to the mountain top is 18 km. William walked up the path at 3 km per hour and walked the same route back at a speed of 4.5 km per hour. What was his average walking speed for the whole journey?

8 There are four whole numbers. The mean of the three smallest numbers is 15 while the mean of the three greatest numbers is 21. The sum of the greatest number and the smallest number is 42. The difference between the other two numbers is 5. Find the 4 whole numbers.

9 There are three types of sweets: Type A is £24 per kg, Type B is £18 per kg and Type C is £12 per kg. If the three types of sweets are mixed together with 2 kg of sweets from Type A, 3 kg from Type B and some sweets from Type C, so that the cost of the mixed sweets will be £16 per kg, how many kilograms of Type C should be put into the mixed sweets?

Chapter 10 test

1 Work these out mentally. Write the answers (in decimals or whole numbers).

(a) $3 - 0.3 \times 10 =$ ☐

(b) $1.8 + 4.2 \div 3 =$ ☐

(c) $5.6 \div 7 \times 8 =$ ☐

(d) $3.81 - (0.81 + 1.5) =$ ☐

(e) $39 \times \frac{8}{13} =$ ☐

(f) $(0.74 - 0.66) \div 8 =$ ☐

(g) $\frac{4}{5} \times \frac{15}{16} - \frac{1}{4} =$ ☐

(h) $\frac{7}{88} \times \frac{11}{35} \times 20 =$ ☐

2 Choose the most appropriate method, mental or written, to find the answer to each calculation.

(a) 17.4×3

(b) $24.32 \div 16$

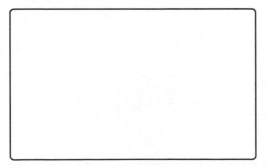

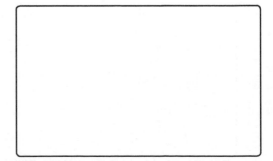

3 Work these out step by step. (Calculate smartly if possible.)

(a) $9.4 \times 3 - 38.4 \div 24$

(b) $(25 + 2.5 + 0.25) \times 4$

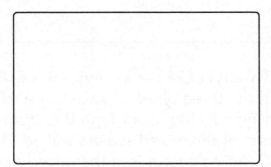

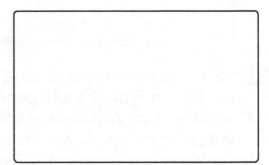

(c) $1600 \div (17 \times 2.5 + 83 \times 2.5)$

(d) $[39 \div (100 - 87) + 6.5] \times 5$

(e) $2\frac{3}{5} - \frac{11}{13} - \left(0.5 + 1\frac{2}{13}\right)$

(f) $4 \times 1\frac{3}{5} - \left(2 \times \frac{11}{15} - \frac{14}{15} \div 7\right)$

4 Complete each statement.

(a) Unit conversion.

(i) $7 \, l \, 20 \, ml = $ _____ l

(ii) $0.023 \, m^2 + 56 \, cm^2 = $ _____ cm^2

(b) In a number, the digits in the hundreds and hundredths places are both 6; 5 is in the ones place. The digits in the other places are zeros. The number is _____ .

(c) The sum of the edge lengths of a cube is 84 cm. The volume of the cube is _____ cm^3 and the total area of its net is _____ cm^2.

(d) If the product of three prime numbers is 105, then the mean of these three numbers is _____ .

(e) If point $P(1, 2)$ is reflected in the x-axis, then the coordinates of its image point are _____ ; if it is reflected in the y-axis, then the coordinates of its image point are _____ .

(f) Translating point P(a, b) first 5 units left and then 2 units down to get

point B, the coordinates of B are _____.

(g) The following line graph shows the numbers of customers visiting a
coffee shop in a week.

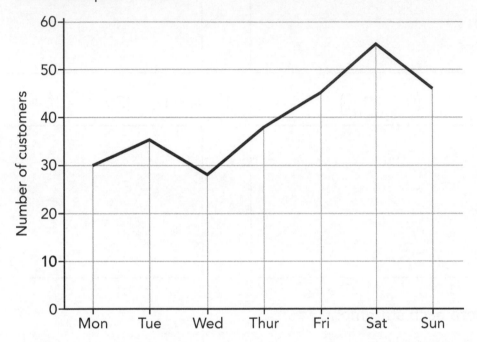

Based on the data shown, the day the shop had the largest number of

customers was _____ and the day the shop had the least

number of customers was _____. The steady increase in

customers is seen from _____ to _____.

(h) If $a : b = c : d$, then $a \times d =$ ☐ $\times$ ☐ .

(i) $15 : 6 =$ ☐ $: 2$; $1.2\,\text{km} : 750\,\text{m} = 1 :$ ☐

5 Multiple choice questions. (For each question, choose the correct answer
and write the letter in the box.)

(a) In the following expressions, the greatest quotient is ☐ .
   A. $13.5 \div 30$ **B.** $1.35 \div 3$ **C.** $13.5 \div 3$ **D.** $1.35 \div 30$

(b) The order of −2.5, 0.25, 0 and −0.25 starting from the least is ☐.

    A. 0.25 > 0 > −0.25 > −2.5   **B.** −0.25 < −2.5 < 0 < 0.25

    C. −2.5 < −0.25 < 0.25 < 0   **D.** −2.5 < −0.25 < 0 < 0.25

(c) Choosing any two different numbers from 1 to 9, there are ☐ possible different sums which are multiples of 5.

    A. 6   **B.** 8   **C.** 9   **D.** 36

(d) If point $A(a, b)$ is in the second quadrant, then $B(-a, -b)$ is in the ☐.

    A. first quadrant   **B.** second quadrant

    C. third quadrant   **D.** fourth quadrant

6 Solve the following equations for x.

(a) $12x + 3.5 = 5.9$

(b) $14x − 8x = 2.7$

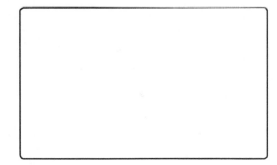

(c) $3(x + 3.2) = 5x$

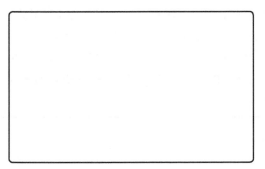

(d) $3.6 − 4(0.9 − x) = 3.2$

7 A road is 360 km long. After a car had travelled on the road for 2.5 hours, it still had 210 km to go. What was the average speed of the car?

8 Matthew walked from his school to a science museum at 80 m per minute. Five minutes later, Mr Lee walked from the school at 100 m per minute to catch up with him. In how many minutes did Mr Lee catch up with Matthew?

9 Twelve workers were binding some books. They could have finished the task on time as planned. However, 3 workers did not turn up for the task for various reasons. Accordingly, the other workers each had to bind 66 more books to finish the task on time. How many books did the workers bind in total?

10 A cuboid water tank is 20 cm long, 18 cm wide and 16 cm high. The water level in the tank is 12 cm high. A piece of iron block of 540 cm³ is put into the water tank and completely immersed in the water. How many centimetres is the water surface away from the top of the container?

11 Place A and Place B are 6 km apart. Tom and Alan started walking from Place A to Place B at the same time. When Alan reached Place B, he immediately went back along the same route. After 50 minutes, Tom and Alan met each other. Alan walked 20 m more per minute on average than Tom. What was Tom's average speed? When they met, how far were they away from Place B?

12 The diagram on the right shows a right-angled triangle *ABC* with *AB* = 3 cm, *BC* = 4 cm and *AC* = 5 cm. Fold side *AB* to side *AC* along line *AD* so that point *B* coincides with point *E*. What is the area of triangle *CDE*?

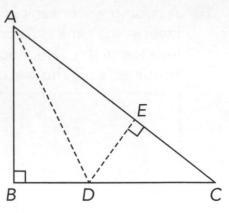

13 800 pupils from a primary school donated their books to a charity. On average, half of the boys donated 8 books each and the other half donated 5 books each, while half of the girls donated 6 books each and the other half donated 7 books each. How many books did all the pupils donate? How many books did each pupil donate on average?

14 A shop bought some pens. The shop found that the profit of selling 20 pens at £10 each is equal to the profit of selling 15 pens at £11 each. What was the cost price of these pens the shop bought?

15 A rectangle has three vertices $A(-8, 3)$, $B(-8, -5)$ and $C(-3, 3)$. Use the coordinate plane below to answer the following questions.

(a) Find its fourth vertex, D, and draw and label the rectangle. What is the area of the rectangle?

(b) Reflect the rectangle in the y-axis and then translate its image 3 units up. Draw and label the rectangle obtained in the new position.

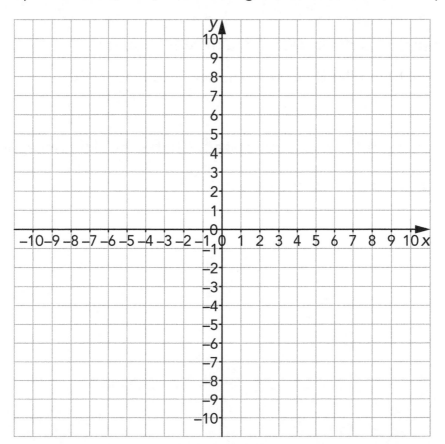

16 The bar chart shows the maximum speeds (correct to the nearest 5 km/h) of five animals. Answer the following questions based on the bar chart.

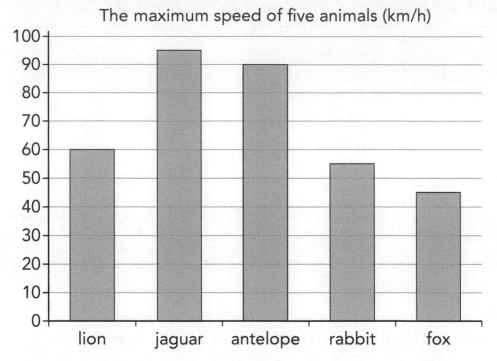

The maximum speed of five animals (km/h)

(a) What animal runs the fastest at the maximum speed? _____

(b) What animal runs the slowest at the maximum speed?

_____

(c) When running at the maximum speed, what is the difference in the

speeds of the lion and the antelope? _____

17 The pie chart shows the percentages of contestants in terms of different ranges of typing speeds in a competition. Answer the following questions based on the chart. (Unit: words per minute)

(a) In what range of typing speeds did the largest number of contestants perform? _____

Percentages of different typing
speeds in competition

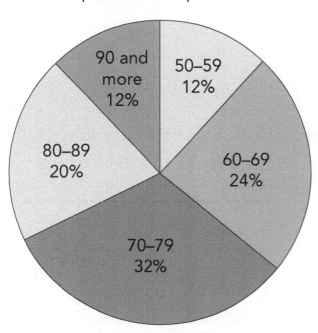

(b) If there were 50 contestants, how many of them performed in the range of 80–89 words per minute? ☐

(c) Can you find the exact mean of the typing speeds of all the contestants based on the pie chart? Why or why not?

_____

_____

(d) Give an estimation of the mean of the typing speeds of all the contestants. Explain how you arrive at your answer.

_____

_____

End of year test

1 Work these out mentally. Write the answers in decimals or fractions in the simplest form. (6%)

(a) $1.8 \times 5 \times 0.2 =$ ☐

(b) $6 \times \frac{1}{2} \div 6 \times \frac{1}{2}$ ☐

(c) $11.6 - 0.2 \times 8 =$ ☐

(d) $2 \times (0.95 - 0.75) + 3.5 \div 7 =$ ☐

(e) $\frac{2}{3} \times \frac{9}{11} - \frac{5}{22} =$ ☐

(f) $\frac{9}{17} \times \frac{51}{126} \times 7 =$ ☐

2 Work these out step by step. (Calculate smartly if possible.) (12%)

(a) $58.475 + 9.89 - 8.475 + 10.11$

(b) $76.48 + 9.6 \div 24 \times 3$

(c) $12.29 \times 28 + 12.29 + 62 \times 12.29$

(d) $\left[0.6 \div \left(\frac{3}{5} + 1\frac{6}{15}\right) + 1.7\right] \times 1.8$

(e) $7\frac{9}{10} - \left(2\frac{2}{5} - 1\frac{4}{15}\right)$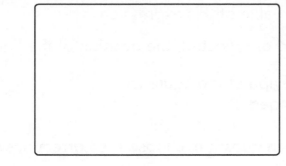

(f) $1\frac{2}{5} \times 25 + 5\frac{1}{4} \div 7$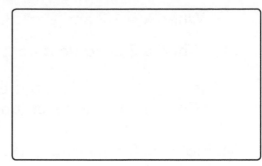

3 Complete each statement. (12%)

(a) 4 lots of $\frac{1}{5}$ equals ⬚.

(b) There are 290 pupils in 8 groups. The mean number of pupils in each group is ⬚.

(c) Simplify the following to the ratios of integers in the simplest form.

(i) $5 : 1\frac{2}{3} : \frac{5}{6} = $ _____

(ii) 20 minutes : $1\frac{1}{3}$ hours = _____

(d) The cost price of a piece of furniture is £3000. If the profit a shop wants to make is 60% of the cost price, then its selling price is

£⬚.

(e) If the number of girls is $\frac{7}{12}$ of the whole class, then the number of boys

is ⬚ of the whole class.

(f) The scale of a map is 1 : 400 000. This means that 5 cm on the map

represents an actual distance of ⬚ km.

(g) The price of a product is first increased by 20% and then reduced by

20%. The current price is ⬚ % of the original price.

(h) Simplifying the expression $3x + 4y - (2x - y)$, it is _____.

When $x = 1.2$ and $y = 1.8$, the value of this expression is ⬜.

(i) When a 2-D figure is translated or reflected, the position of the figure

is _____, but the shape of the figure is _____.
(Choose: 'changed' or 'unchanged'.)

4 Multiple choice questions. (For each question, choose the correct answer and write the letter in the box.) (9%)

(a) The correct statement of the following is ⬜.

    A. There are infinitely many factors of 8.

    B. There is no common factor for any two prime numbers.

    C. The common factor for all positive integers is 1.

    D. The least natural number that can be divided by both 4 and 6 is 24.

(b) If $\frac{5}{x+2}$ is an improper fraction and x is a positive integer, then there

are only ⬜ possible integers for the value of x.

    A. 1   **B.** 2   **C.** 3   **D.** 4

(c) Two pairs of numbers for x and y that both satisfy the equation

$2x + 5y = 7$ are ⬜.

    A. $x = 1, y = 1$ and $x = 0, y = 1.5$

    B. $x = 1, y = 1$ and $x = 0.5, y = 1.2$

    C. $x = 3.5, y = 0$ and $x = 2, y = 1$

    D. $x = 0, y = 1.4$ and $x = 1, y = 2$

(d) It takes 2 hours by bus from an airport to the city centre and only 7 minutes by high speed train. The ratio of the speeds of these two

methods of transportation is ⬜.

    A. $120:7$ **B.** $7:120$ **C.** $2:1$ **D.** $1:2$

(e) The expression with the same result as 3.75×160 is ⬜.

    A. 0.375×16 **B.** 375×0.16 **C.** 375×1.6 **D.** 37.5×1.6

(f) Put 3.45, 3.4̇, 3.54 and 3.448 in order, starting from the least.

The second number is ☐.

A. 3.45   **B.** 3.4̇   **C.** 3.54   **D.** 3.448

(g) There are 60 spare parts in a box. They weigh 2 g, 5 g and 9 g. The number of spare parts of 2 g is twice that of 5 g, which is three times that of 9 g. The total weight of the spare parts is ☐.

A. 200 g   **B.** 216 g   **C.** 225 g   **D.** 270 g

(h) If point $P(-1, -1)$ is reflected in the x-axis, then the coordinates of its image P' are ☐.

A. $(-1, 0)$ **B.** $(0, -1)$ **C.** $(-1, 1)$ **D.** $(1, -1)$

(i) If point $P(a, b)$ is translated 3 units right and 2 units down, then the coordinates of its image P' are ☐.

A. $(a + 3, b + 2)$ **B.** $(a - 3, b - 2)$

C. $(a + 3, b - 2)$ **D.** $(a - 3, b + 2)$

5 Solve the equations for x. (8%)

(a) $2 \div x = 5$

(b) $48x - 70 = 50$

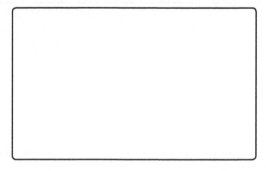

(c) $3(x + 3.5) = 13.5$

(d) $38x + (16 + 12x) = 24$

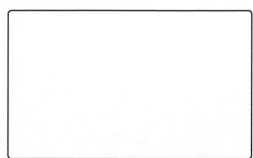

6 Given $x : \frac{1}{2} = (1 - 4x) : 7$, find the value of x. (2%)

7 Given $x : y = \frac{1}{2} : \frac{1}{3}$, $x : z = \frac{1}{4} : \frac{1}{5}$, find $x : y : z$. (2%)

8 James walked along a 50 m-long path four times. It took 76 steps the first time, 78 steps the second time, 80 steps the third time, and 78 steps the fourth time. How many metres is the average length of each of James' steps? (Keep your answer to two decimal places.) (4%)

9 A software company was making one batch of software. It was scheduled to make 120 pieces each day and finish the job in 30 days. With better efficiency, the job was actually done 5 days ahead of the schedule. Compared with the original schedule, how many more pieces of software were made each day? (4%)

10 An orchard has 200 peach trees and pear trees in total. The number of peach trees is 10 less than twice the number of pear trees. How many pear trees and peach trees are there in the orchard, respectively? (Hint: Use an equation to solve the problem.) (4%)

11 Look at the diagram. The perimeter of the square is 32 cm. A is the midpoint of the side of the square. Find the area of quadrilateral ABCD. (4%)

12 The diagram shows a right-angled triangle *ABC*. The perimeter of the triangle is 60 cm, *BC* = 15 cm and *AC* = 20 cm. What is the height of the triangle from the vertex *C* to the side *AB*? Show your working. (4%)

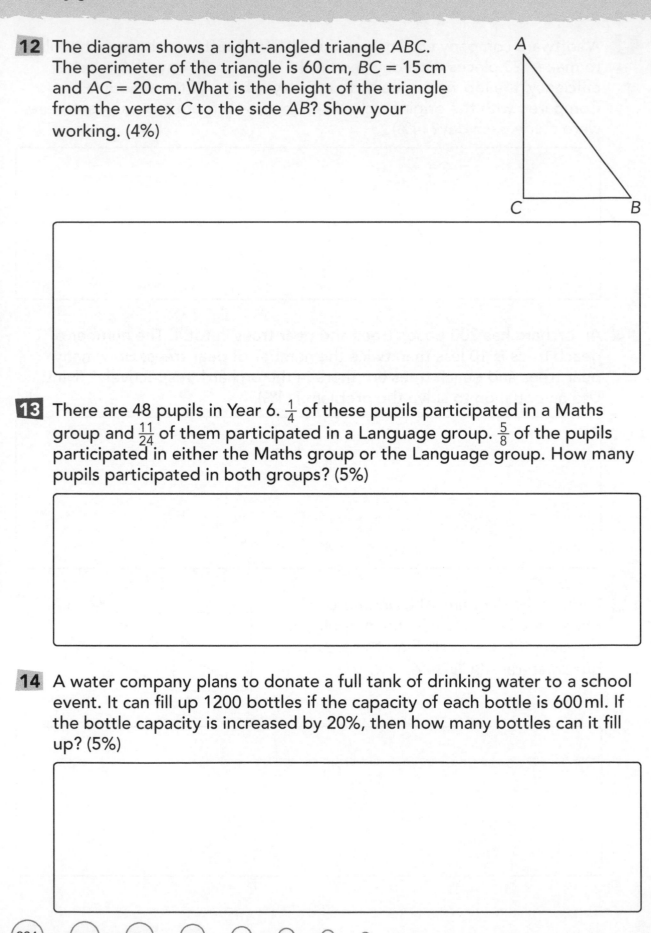

13 There are 48 pupils in Year 6. $\frac{1}{4}$ of these pupils participated in a Maths group and $\frac{11}{24}$ of them participated in a Language group. $\frac{5}{8}$ of the pupils participated in either the Maths group or the Language group. How many pupils participated in both groups? (5%)

14 A water company plans to donate a full tank of drinking water to a school event. It can fill up 1200 bottles if the capacity of each bottle is 600 ml. If the bottle capacity is increased by 20%, then how many bottles can it fill up? (5%)

15 A dress suit was originally priced at £750. During the first promotion, the shop offered 20% off the original price and made a profit of £90 per suit. To sell the suits faster, the shop plans to have a further promotion, what is the maximum additional discount it can offer without making a loss? (5%)

16 The table below shows the ticket prices of three sports events. A company purchased 100 tickets for the events. The pie chart shows the distribution of the tickets bought for the different sports events. (6%)

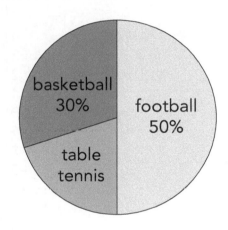

Sport event	Ticket price
football	£100
basketball	£80
table tennis	£50

Using the information shown above, answer the following questions.

(a) ☐ % of the total tickets purchased were for table tennis.

(b) The number of tickets purchased for football was ☐ .

(c) Of all the money spent on the tickets, what fraction of it was spent on the tickets for table tennis? ☐

17 A square has three vertices: $A(1, 3)$, $B(6, 3)$ and $C(6, -2)$.

Use the coordinate plane below to answer the following. (6%)

(a) Find its fourth vertex, D, and draw and label the square.

(b) Which quadrant(s) is the square in? What is its area?

(c) Reflect the square in the y-axis and then translate its image 3 units up. Draw and label the square obtained in the new position.

Which quadrant is the new square in now? _____

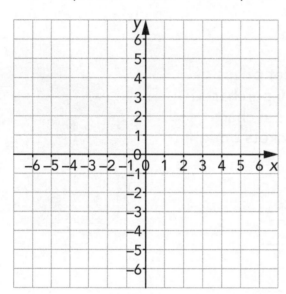